Industry 4.0 Topical Conference

Topical Conference at the 2021 AIChE Spring Meeting and 17th Global Congress on Process Safety

Online
18 - 23 April 2021

ISBN: 978-1-7138-5327-5

Printed from e-media with permission by:

Curran Associates, Inc.
57 Morehouse Lane
Red Hook, NY 12571

CURRAN ASSOCIATES INC. proceedings .com

Some format issues inherent in the e-media version may also appear in this print version.

AIChE
120 Wall Street, FL 23
New York, NY 10005-4020

Phone: (800) 242-4363
Fax: (203) 775-5177

www.aiche.org

Additional copies of this publication are available from:

Curran Associates, Inc.
57 Morehouse Lane
Red Hook, NY 12571 USA
Phone: 845-758-0400
Fax: 845-758-2633
Email: curran@proceedings.com
Web: www.proceedings.com

TABLE OF CONTENTS

620660 The Digital Transformation Promises to Change Innovation in the Materials Industry. Will It?

Abstract:

The industry is challenged to innovate faster than ever while also delivering sustainability improvements that require profound change through the value chain. Simultaneously with that pressure, digitalization and AI are advertised as revolutions that has caused many to fear becoming obsolete. This dichotomy of extreme pressure and emerging, revolutionary technologies is causing fast, significant change in the industry and the innovation process. In this talk, I would like to share my view on what this impact will be, what risks exist, and what, in my view, will drive long-term success.

620661 Digital Twins & Smart Manufacturing –*Where Data Analytics Meet Modeling & Optimization*

Abstract:

Modeling and Optimization constitute core competencies of process systems and chemical engineering fields, and play pivotal role in the analysis, simulation, design and operational activities that we perform daily in order to improve the performance and sustainability of energy and manufacturing processes and supply chains. The re-emergence of AI and Big Data revolution that we are all witnessing at the moment, also provide an opportunity and a platform toward real-time decision-making. While often perceived as different fields, in the context of Smart Manufacturing, Data Analytics and Modeling/Optimization can in fact be viewed as 'two sides of the same coin'.

In this lecture, we will attempt to elucidate this argument through the lenses of ***hybrid modeling and digital twins***, which bring together the advantages and power of high-fidelity modeling & optimization and AI/machine learning technologies and tools, and their synergy. A central theme will be that the evolution/revolution in data-driven methods in the context of process modeling and automation is becoming a critical component in advancing the energy & manufacturing transition agenda toward clean energy, increased process/energy efficiency, sustainability and resilience, and ultimately economic and societal prosperity.

620586 Navigating the Growing World of Analytics: Separating the Right Data from Big Data

Abstract:

Manufacturing generates more data than another sector, but as companies that successfully navigate digital transformation know, data alone isn't enough.

Chief Technology Officer Peter Zornio will share Emerson's philosophy on big data and reveal how operational analytics technologies are critical to tackle complex challenges and optimize operations to drive measurable performance improvements.

620560 Delivering Value through Artificial and Virtual reality at Pfizer

Abstract:

At Pfizer, a multi-year digital transformation is driving improvements in cycle time, yield, and right-first-time quality. Mike Tomasco, Vice President of Pfizer Digital Manufacturing will share the digital transformation journey of Pfizer's manufacturing facilities.

619954 In-House Development of Image Analysis Solutions for the Chemical Industry

Abstract:

Image analysis is a fundamental aspect of modern research, as many laboratory test methods consist of visual output that requires manual inspection or rating to assess product performance. The breadth of image analysis applications is expansive across a number of industries, including, for example, quantitative assessments (such as object or pattern recognition, color evolution, defect analysis, size determination), time-lapsed visual comparisons, and qualitative assessments (such as aesthetic judgements or ratings comparing relative performance or past experience). The automation of image processing provides efficiency gains and standardization for these methods, ultimately improving the consistency and quality of data in a research setting. Experience-based visual methods in particular suffer from low efficiency, lack of consistency, and poor reproducibility, which can be substantially improved with automated image analysis. Tests and results can be stored digitally for better knowledge retention and access across an organization. Furthermore, the available information in the database can be built-up over time to accommodate big data analytics and in-depth statistical analyses in the future. As an industry leader, Dow has developed in-house expertise and methodologies that leverage digital image processing across many businesses.

In this talk, we will demonstrate the scope of image analysis problems and discuss one specific tool development in detail: microscope image stitching of film samples in the plastics industry. We will highlight the context of the test method, technical challenges, the reason to choose in-house development, and the strategies and challenges of method development and validation.

619943 Materials Modeling - from the Molecule to the Final Product

Abstract:

Chemical & process companies cater to a wide range of industries such as aerospace, automotive and consumer products. Chemical products from the process industries need to satisfy the performance and safety requirements from their customers. This entails that the materials produced by the chemical industry needs to be tightly linked to the product requirements of their customers.

Designing a plane wing, a windmill turbine or a car bumper cannot overlook the impact of material choice on the final property of the part. Materials are getting more and more complex, and specific fine tuning is necessary to achieve the expected results. This is particularly important when it comes to polymers or composite materials as the variety of chemical modifications that can be performed on the row materials are basically limitless. This calls for a better integration of different field of science from chemistry to chemical engineering and processing and design, geometry or material science.

In this talk we will give a broad overview of the most recent applications of modeling, simulation and data science to predict the final mechanical properties of parts and products. Starting at the chemical level and going up the scales up to the processing of the part and its manufacturing. We will also present some of the tools that allow researchers and designer to do so and propose our vision on the "missing links" that should be the focus of researchers for the coming years to facilitate a full multiscale modeling of polymer and composite parts.

618916 A Framework to Identify Weak Signals of Process Incidents Using Fram and Machine Learning Algorithms

Abstract:

Incidents in process industries still occur globally even though professionals put many efforts into improving process safety over the decades. Although randomness may be an element, incidents are often not chance-events in that these suddenly occur out of nowhere. They do follow incubation periods where weak signals exist a long time before the incidents occur and these signals need to be recognized and resolved as early as possible to prevent the incidents. Weak signals emerge as complex interactions among performance variabilities of human, organization and technological functions, therefore their identification could be intellectually unmanageable. In the study, a framework to identify weak signals was developed and consisted of a system-based hazard identification method (i.e., Functional Resonance Analysis Method) and machine learning models. Its application in a batch polymerization process was implemented for illustration. The FRAM was first applied to understand the interactions in the system and determine what information would be needed for weak signal identification. The information was generated by a digital twin, which was developed based on the FRAM to simulate possible function interactions in the batch process. Finally, Random Forest was developed to predict the probability of unexpected temperature deviation given a set of potential weak signals, and Decision Tree was developed to identify the corresponding combination of weak signals, which can proactively predict the unexpected scenario.

620031 A "People-Focused" Approach for Manufacturing in Industrial Internet of Things (IIoT) Environments Boosts Productivity

Abstract:

The industrial internet of things (IIoT) has brought increasing emphasis on interconnecting sensors and devices in manufacturing facilities. Now applications utilizing cloud computing and AI are helping organizations leverage their IIoT infrastructure. This has help to move IIoT beyond the confines of IT to the shop floor, all to optimize efficiencies and productivity.

Even though this transformation is happening rapidly, it is vital not to overlook the *fact that manufacturing still needs people* (my, hasn't COVID proven this?). While IIoT enables faster and better reactions to operating conditions, even better trained staff are needed to understand and control these systems and to adapt them to changes more rapidly. Thus a "people-focused" IIoT enables industrial enterprises to harness true workforce know-how to maximize plant operations by increasing production efficiency which, most importantly, ensures employee safety.

Plant Process Management (PPM) systems help manage, monitor and optimize plants in addition to work activities, operations and production assets. They give a real-time view of work progress and track status of any incidents or outliers to routine operations. Current employees working in shift manufacturing need to be made aware of any issues that may have occurred in the previous shift, which isn't always easy. Poor communication can directly impact safety, quality and productivity.

PPM delivers on the requirements for all plant operations, regardless of the use of automation, IoT instrumentation, machine learning (ML) or predictive maintenance. Knowledge can be gained from any human contact point in a process. Consistent capture of team-to-team communication is critical, especially with changing work patterns due to working remotely or social distancing. For example, COVID-19 has put a strain on conducting safe operations but fortunately PPM is one solution that can safeguard manufacturers and further optimize their operations.

For most manufacturers, IT and OT represent sets of separate and distinct information silos, hindering the progress of IoT. PPM solutions help bridge this gap by allowing teams to seamlessly interact with OT data and IT applications, for example.

In a survey conducted by 451 Research, part of S&P Global Market Intelligence, more than 50% of OT respondents surveyed confided that they seldom work together with their IT team. Of those surveyed, only about 20% of IT leaders collaborate with OT. IT teams believe the key problem lies in the inability to share captured data where OT teams feel reconciling shift data with machine logs poses the biggest challenge. However, both IT/OT participants surveyed agreed the following points are a top priority:

- Capturing and sharing best practices for employee safety
- Implementing real-time incident tracking to conduct corrective action
- More cohesive collaboration and communication among IT/OT teams
- Automated interactive production loss accounting

According to research findings, a focus on worker safety is an important driver for changing the status quo. Without repeatable procedures in place where human communication occurs, a disconnect can occur where bad incidents could happen. As equipment and resources are run and maintained by a human workforce, the correct tooling and equipment, including those enabling clear communication and collaboration around details are essential.

It is possible to digitally transform operations with PPM, without a full IIoT roll-out. Consider quick wins such as digitizing shift handovers, safety and compliance tracking, and corrective maintenance planning. Or what about replacing whiteboards, ad hoc conversations and checklist sheets with an electronic system that anyone can access, anywhere? Once digitized with PPM, these common elements of everyday shift work can easily take advantage of future IIoT capabilities and become powerful assets in themselves to the production operation.

However, if the workforce is not taken into account in your digital transformation, potential benefits can be lost. In addition, as older workers retire and are not replaced as quickly as they leave, vital knowledge from their years of experience can disappear without a digital PPM-based knowledge base. Furthermore, the COVID-19 pandemic, as well as the potential for future crises (whether global or local), highlights the necessity for digital tools that help teams communicate, coordinate, and co-operate – wherever they happen to be.

Consider PPM as the next step in your path to digital transformation.

620008 Open Process Automation: Enabling New Technologies for Manufacturing

Abstract:

The move to industrial control systems based upon open standards has the potential to bring new and emerging technologies resident in the edge, the cloud, and anywhere in-between into manufacturing. The work being progressed by members of the Open Process Automation Forum is leading the transition to a standards-based, open, secure, and interoperable process control architecture. The ideals of Industry 4.0 are significantly aligned with those of Open Process Automation, which can be considered a means for deployment. Systems built from components aligned with the Open Process Automation Standard (O-PAS™) and its principles allow for significant value capture by integrating emerging technologies such as AI, machine learning, and advanced data analytics. Much of this value comes through leveraging the massive amount of complex and heterogeneous data generated by, or ultimately connected to industrial control systems and using it to solve complex business and engineering problems. This talk will describe ExxonMobil's vision for Open Process Automation and highlight the milestones reached on this journey to date. This will include a description of ExxonMobil's test bed facility and how it is being used to assess the completeness and practicality of elements of the standard in a real world setting and to achieve technical readiness for specific commercially available components to be used in permanent installations.

620167 Unlocking Value through Better Quantification of Uncertainty in Asset Utilization Loss Using a Data-Driven Stochastic Approach

Abstract:

Motivated by the extensive labor and capital costs and potential loss of revenue at stake, optimization tools have been successfully developed to support turnaround planning for maximizing the value of manufacturing asset operations. For instance, Amaran et al. [1-2] consider medium and long-term turnaround planning in integrated sites using mathematical programming approaches to balance different trade-offs.

Some of the major drivers for a turnaround include asset performance renewal, regulatory inspections, and capital projects. Of these drivers, asset renewal presents a more flexible window of opportunity to unlock operational value through optimal turnaround interval planning. However, evaluating this opportunity requires characterization of Asset Utilization (AU) loss over time to subsequently incorporate into the existing planning framework.

Asset-level modeling based on individual equipment reliability analysis can be overly complex and expensive to maintain from a data collection and modeling standpoint. Resorting to higher level models which are less complex, may result, at times, in multiple individual effects getting aggregated. This can lead to a set of categories with observations lacking a discernible trend with respect to the actual root-cause factors. In such cases, Hidden Markov Models (HMM) [3] can be useful in modeling the uncertainty around the observations by assuming that the system transitions through non-observable or hidden states.

In this work, we consider a data-driven stochastic approach to model AU loss for an asset based on conceptualized production loss historian data. The conceptual historian tracks asset utilization loss over a set of equipment or process categories over time. We build HMM models for a subset of the categories without a discernable trend using the historical data as observations. Further, this approach combined with Monte Carlo simulations allows us not to be trapped by overly conservative predictions in AU losses. We illustrate the application of this approach within a turnaround planning model.

References:

[1] Amaran, S.; Sahinidis, N.V.; Sharda, B.; Morrison, M.; Bury, S. J.; Miller, S.; Wassick, J. M. Long-term turnaround planning for integrated chemical sites, *Computers & Chemical Engineering*, 2015, 72, pp. 145-158.

[2] Amaran, S.; Zhang, T; Sahinidis, N. V.; Sharda, B.; Bury, S. J. Medium-term maintenance turnaround planning under uncertainty for integrated chemical sites, *Computers & Chemical Engineering*, 2016, 84, pp. 422-433.

[3] Chen, Z.; Li, Y.; Xia, T.; Pan, E Hidden Markov model with auto-correlated observations for remaining useful life prediction and optimal maintenance policy, *Reliability Engineering & System Safety*, 2019, 184, pp. 123-136.

619993 Data-Driven Approaches for Prediction of Industrial Catalyst Deactivation

Abstract:

Data-Driven Approaches for Prediction of Industrial Catalyst Deactivation

Adrien Spagnol[1], Kevin Jaunatre[1], Gabriel Ducret[2], Pascal Duchene[1], Denis Guillaume[1], Vinay Jain[1] and Benoit Celse[1]

1 IFP Energies Nouvelles, Rond-Point de l'Echangeur de Solaize, 69360 Solaize, France

2 IFP Energies Nouvelles, 1-4 Avenue du Bois Préau, 92852 Rueil-Malmaison, France

**E-mail: benoit.celse@ifpen.fr*

HIGHLIGHTS

- Comparison between different data-driven approaches on industrial dataset.
- Linear regression techniques provide easy-to-explain models.
- Validation of the selected models on chosen cycles, with discussion regarding the possible limitations.

INTRODUCTION

More than 95% of all the chemical products are nowadays produced using a catalyst, which shows the high importance of catalysis for all value chains in the chemical industry, especially in petroleum refineries. In that scope, catalysis deactivation as an interdisciplinary technology field is of strategic socio-economic importance, since it is a considerable and enduring problem when operating industrial catalytic processes. Thus, being able to predict the deactivation of the catalysts is a primary concern.

From the statistical point of view, correctly predicting the deactivation phenomenon in an industrial environment is a challenging task due complex chemical processes and affecting factors such as reactor configuration, quality of feed oils, the catalyst degradation, and greatly varying operating conditions.

Based on collected data from different petrochemical refineries with varying capacities, feedstock's, and operating conditions, see Table 1 below, the main objective of this work is to build a predictive model of the catalyst deactivation for Vacuum Gas Oil (VGO) Hydrotreatment (HDT) in order to correctly forecast the cycle duration.

Number of refineries	Number of cycles	Average length (in days)
12	21	830

Table 1. Summary of the collected data from petrochemical refineries

MATERIAL & METHODS

Two kind of models are tailored depending on different scenarios. In a first place, the Design configuration, where the goal is to predict the duration of a cycle based on a launch setup, assumed constant over the whole life of the catalyst. In a second place, a What-if configuration, whose goal is to forecast future deactivation,

Prior to any modeling, the preprocessing of the data is detailed, in order to underline the importance of the quality of the used datasets, correct existing caveats from measured information and propose improvements. Different type of models are discussed in this work, either regression-based approaches, which build a relationship between historical data and the catalyst deactivation rate, or time-series strategies, since the catalyst deactivation phenomenon has a time-dependency.

For regression strategies, we explore many available options from the literature, from multilinear regression, possibly including a penalization term [1][2], to nonlinear approaches [3,4]. For time-series techniques, we test simpler approaches [5] and complex models including regressors [6].

RESULTS & DISCUSSIONS

All strategies are discussed, listing the different pros and cons of each. A comparison is done using a calibration-validation process in order to rank each technique based on their quality of prediction, for both scenarios.

Best selected approaches are used on experimental cycles to assess the benefits of being able to predict catalyst deactivation, compared to the classical approach used on refineries based on a straightforward time extrapolation of the weighted average bed temperature (WABT).

Furthermore, building a input-output model allows to optimize the economics of the refineries by offering insights on how to conduct the end of a given cycle, based on the predictions (i.e. showing if it is reasonable to consider different feeds, possibly harder to treat).

KEYWORDS

Catalyst deactivation, regression techniques, time-series approaches, Vacuum Gas Oil Hydrotreatment

BIBLIOGRAPHY

1. Robert Tibshirani, "Regression shrinkage and selection via the lasso," *Journal of the Royal Statistical Society: Series B (Methodological)*, vol. 58, no. 1, pp. 267-288, 1996.

2. Hui Zou and Trevor Hastie, "Regularization and variable selection via the elastic net," *Journal of the Royal Statistical Society: Series B (Statistical Methodology)*, vol. 67, no. 2, pp. 301-320, 2005.

3. Alex J. Smola and Bernhard Schölkopf, "A tutorial on support vector regression," *Statistics and computing*, vol. 14, no. 3, pp. 199-222, 2004.

4. Leo Breiman (2001). Random forests. *Machine learning*, 45(1), 5-32.

5. Everette. S. Gardner, "Exponential smoothing: The state of the art," *Journal of Forecasting*, vol. 4,

no. 1, pp. 1–28, 1985.

6. George Athanasopoulos Rob J Hyndman, *Forecasting: Principles and Practice*, OTexts, Ed., 2018.

619992 Auto Plot Pathfinder : Can Multi-Purpose Optimization Generate a Plant Layout Comparable to a Veteran Engineer ?

Abstract:

JGC has been developing a generative design method for plant layout -what we call Auto Plot PATHFINDER-, applying satisficing trade-off method implemented with genetic algorithm and JGC expert knowledge. It instantly generates numerous numbers of layout designs that either an engineer or an algorithm alone cannot generate.

The exploration in the generative designs is formulated as an area-constrained knapsack problem, in which process units are allocated to strike the trade-off balance between minimizing a piping cost and minimizing a layout area. The method is developed analogically based on the JGC experts' way of plant layout design. It consists of a rule portion such as complying with maintainability, safety, and regulations, and an optimization portion such as pursuing economic efficiency and competitive advantage. The rules are developed with a design structure matrix that formulates expert knowledge, while the genetic algorithm works for the optimization portion. Setting the aspiration level of piping cost and layout area enables users like piping engineers to control the exploration space. Furthermore, the objective function has such flexibility that the users can embed their domain knowledge into the objective function. As characterized above, the practical generative design is realized by a hybrid optimization method of mathematical techniques and engineering expert knowledge.

A proof-of-concept trial where Auto Plot PATHFINDER has been demonstrated with the past design projects has experimentally proven it broadly applicable to plant design. We have confirmed users can manipulate the exploration space and the optimization result by setting the aspiration level. A Pareto-frontier of the 2-dimension case that comprises piping costs and layout area shows the plant layout generated and the asymptotic behavior under minimum piping cost and minimum layout area are found consistent with our experts' intuition and insight, suggesting that the expert knowledge is successfully embedded into the algorithm. In addition, it surprisingly reproduces the creativity of veteran engineers. The analysis of the objective function shows us the very small techniques that veteran engineers often utilize make such a huge contribution to reducing piping costs as approximately 40% reduced. Auto Plot PATHFINDER can help us with discovering what makes a great plant layout great.

619948 Enabling Digital Transformation: It's about More Than the Analytics

Abstract:

Enabling Digital Transformation: It's about more than the Analytics

Over the past 5 years, the manufacturing industry has come to understand the potential impact of and the value of data analytics in the process environment. What followed was a deluge of vendors, each shilling their particular flavor of analytics – machine learning, multivariate, artificial intelligence, etc. In spite of the multitude of analytics options, most manufacturers continue to struggle with fully leveraging analytics output to deliver actionable insights in any scalable and sustainable way. As the speed of and emphasis on digital transformation increases, those struggles have been exacerbated.

Despite the growth of investments in individual analytics technologies, manufacturers are only experiencing marginal gains in their ability to analyze, understand and implement insights generated from the data. The manufacturers setting the pace in the industry today understand the need to move beyond the simple application of data analytics. They are instead starting to look at *enabling* their organizations with analytics knowledge.

Through analytics knowledge enablement, those insights can be leveraged anywhere in the company to anticipate and solve potential process upsets faster and with greater predictability. These leading manufacturers understand that generating the initial signal with analytics is only the first step. The larger value comes from the ability to then harvest and harness the knowledge generated across the organization.

This session will:

- Highlight the critical differences between simple analytics usage and manufacturing analytics knowledge enablement;
- Demonstrate how manufacturing analytics knowledge enablement delivers competitive advantage;
- And the requirements to deliver scalable and sustainable manufacturing analytics knowledge enablement.

619075 Leveraging Digital Advances to Optimize Maintenance Activities

Abstract:

This presentation discusses how digital advances can optimize maintenance activities, and the performance of an overall operation, through specific use cases. The cases review how mobile devices such as tablets can be linked remotely to monitoring and control facilities from multiple locations. Maintenance crews collaborate in real time with experts—through audio, video, or chat—to improve asset management and maintenance planning. Smart apps help integrate tablets into the plant environment, enabling access to asset-specific information, including past maintenance and recent operations data. AR advances enhance the experience, allowing the remote expert to overlay repair instructions onto the shared image.

Digitization solutions such as these also draw on advances of the digital twin concept. Because a wealth of equipment and operations data is available digitally, it can be easily downloaded to the maintenance person's tablet for immediate use. This literally gives maintenance workers everything they need to resolve an equipment problem, right in the palm of their hands. Remote linking between this tablet and the expert's onshore desktop affords much faster service and a lower risk of lengthy, unplanned shutdowns.

During this presentation you will gain the following insights:

- A general understanding of key digitalization topics such as the digital twin, 3D, simulation as well as innovative ways of collecting data
- How can information and data be used to more productively operate and maintain a plant?
- Learn how maintenance can be done more efficiently with digital technologies

621794 Sponsored Technology Workshop - Three Myths About Enabling High Performance Teams with Digital Technology - Eschbach

Abstract:

Digitalization inevitably changes team dynamics. Companies often focus on technology and fail to understand how teams actually work. In this high-energy and interactive work session, Eschbach founder & CEO Andreas Eschbach and Industry Risk consultant Andy Brazier will reveal three mistakes process manufacturers make when trying to digitize their industrial workforce and how understanding human factors will help you to avoid them.

618962 Maximizing Batch Yield in the Presence of Upstream Disturbances with Multivariate Modeling and Numerical Optimization

Abstract:

Tightly controlled batch processes typically aim to replicate "optimized" results by restricting variation from an established operating profile. However, this operating strategy, by design, limits the ability to adapt to process disturbances. Even a small relaxation of control limits on key variables can make a significant impact on the ability to achieve the desired final batch quality in the presence of a disturbance.

This presentation will include a batch reactor case study that demonstrates a 40% predicted improvement in process yield by adjusting the pre-established operating profile to account for a feed quality disturbance using multivariate analysis (MVA) and numerical optimization.

MVA is a proven method for analyzing and interpreting diverse industrial Big Data. MVA has been successfully applied to data from batch processes for a variety of goals, such as improving yield and product quality, and reducing operating costs. Numerical optimization in this case study is a true and formal mathematical optimization, in which the solution is distinctly different from the pre-established operating profile coined as "optimized". In many cases, the pre-established profile is often simply the historical batch that produced the best quality (or yield in this case) and is used as an initial feasible starting point.

In this application, a PCA model is developed to reduce the dataset from the upstream process to a small number of latent variables that are easily monitored on a score plot for any feed quality disturbance to the downstream batch reactor. A multi-block PLS is then developed to correlate upstream conditions (from the PCA model results) and batch reactor process conditions to final batch yield.

Next, numerical optimization is performed to calculate the batch operating profiles that will maximize batch yield. The Sequential Least Squares Programming (SLSQP) algorithm included in the SciPy Optimization package in Python is used to solve the nonlinear programming (NLP) problem. Extrapolation in the optimization step is constrained through the use of the nonlinear SPE and HT2 terms from the PLS model, which ensure that resulting profiles are within historical operating limits.

Finally, the optimization results are compared to the pre-established operating profile results.

This presentation will introduce the concepts of MVA in the context of the case study, and will discuss typical challenges and dataset limitations for tightly controlled batch processes.

618576 How to Apply AI at Scale at a Chemicals Plant

Abstract:

Successfully deploying your first AI model is a critical initial step towards a data-driven operations floor. Once you have developed and deployed your first successful model into production (live streaming data in and live predictions out) you'll want to do more. Different models, different systems, in different parts of the plant.

The issue becomes how do you add the second, third and tenth model without incurring a huge amount of IT resources and long cycles of IT development. Plus, how do you keep it all running.

Scaling AI across multiple processes brings new complexities that must be mitigated to achieve true automation. In this session learn about the key considerations and requirements to scaling your AI strategy, including how to connect the different sources of data; managing dozens of models concurrently; and monitoring AI for success.

617185 Industry 4.0 and Control Valve Analytics

Abstract:

A process plant may have hundreds or even thousands of control valves, and 30% or more may be facing physical or mechanical performance issues. This practical session shows attendees how to leverage control valve analytics to deliver bottom-line value. Using the constructs of Industry 4.0, attendees will learn about many available analytics, develop accurate diagnoses, and use that information to effect significant changes to reliability, production, quality, and safety.

Microsoft defines 5 necessary layers for the internet of things: Smart devices, connectivity, history, analytics, and presentation. When it comes to control valves, these 5 layers have been in place for decades. So, manufacturers are well-positioned to gain value from an existing installed base, without investing significant capital.

Even though the raw technology of digital analytics for control valves has been available for more than 20 years, these tools have been inconsistently applied. There have been significant barriers to data acquisition and storage, inconsistencies between vendor tools, and inconsistent practices and procedures to leverage valve data for corporate gain.

Whether a valve is equipped with a smart positioner or not, it is possible to accurately diagnose valve issues. Some techniques are quite simple, and others more sophisticated, using pattern recognition to pinpoint issues. Software tools have simplified the process, and made these capabilities readily accessible remotely, so analysis can be done from the plant floor, or from a home office.

This paper covers available tools and techniques to acquire control valve data, to develop meaningful analytics and diagnostics, and to implement a structured program for managing a fleet of control valves. Specific examples will show how to automatically detect valve issues, how to automatically prioritize, and how to manage resolution of these issues.

618850 Benefits of a Digital Operations Platform

Abstract:

Topic:

This abstract is about the industry benefits of a digital operations platform. I will speak to the commercial and operational benefits of applying and end-to-end approach reaching from digital twin project data to digital operations and connecting to digital upkeep and day-to-day maintenance.

A platform such as this extends across:

-> first principles based digital process twin model

-> emulation of process control technology for validating control philosophy

-> digital automation twin engineering connected to the previous

-> virtual commissioning of the process and control strategies

-> operator training connected to 3D digital twin, digital process twin and digital automation twin

-> real-time optimization of chemical process and reactions based on the previous

-> machine learning and artificial intelligence for health monitoring of control assets and process equipment

Digital Operations is a single platform with seamless plug-and-play technology modules that delivers business capabilities relevant for chemical industry, without any gaps between different teams, stakeholders or hand-over situations.

Overview:

Digital operations are part of a digital thread that reaches from research and development to project design to operations and finally to electronic marketplaces.

Digital transformation in the chemical industry is complete once the digital thread that is relevant has been completed and covers the entire value chain that applies to the lifecycle of a chemical plant or fleet of plants.

Digital operations consist of receiving the deliverables of a digital design, and using that information for training, construction, commissioning, operation, and optimization of capital assets. Digital design deliverables are typically managed starting from requirements in an enterprise-level digital collaboration platform and authored in 1D, 2D and 3D tools that are accessed from the construction site before handed over to operations. Digital plant maintenance and management of change complete digital operations and digital designs. All three areas mentioned here must be connected to enterprise-level resource planning and in the end to commercial entities i.e. customers.

Benefits:

The value of digitalization and the degree in which it is adopted in plant operations, will significantly benefit all phases of the plant lifecycle, focusing on:

- Maximum levels of personnel and process safety
- Total cost of ownership, operation, maintenance, energy
- Highest yields and throughput
- Improve equipment uptime
- Skilled workforce deployment by leveraging operator training systems
- Reduce complexity and the number of manual tasks
- Single access point to all process, project, and maintenance information
- Accelerate optimization by means of model-and-data-based simulation
- Reduce project duration, complexity, and effort by eliminating custom integration
- Rapid deployment supported by virtual commissioning functionality

Main beneficiaries related to digital operations can be assumed to be the plant manager, plant engineers, plant process operators, reliability engineers, HSE, production scheduling, plant maintenance technicians, plant turnaround coordinators, 3rd party vendor staff, on-call staff, cybersecurity manager, corporate process automation group and any commercial stake-holders.

Relevance:

The end-to-end digital operations platform described in this abstract is compatible with such programs as Manufacturing USA and Industrie 4.0 that describe digital transformation and digital enterprises in the context of industrial activities.

619280 The Refinery of the Future

Abstract:

Delek US, an energy company focused on the oil & gas market within the United States, is transforming their business using innovative technologies. This presentation will highlight their vision of "The Refinery of the Future" by how they are managing Risk, AI and Cyber Systems to make their refineries safer, more efficient and sustainable.

This talk will focus on integrating new technologies to provide a comprehensive real-time view for plant operations. The Command Center creates a more responsive environment where the "Connected Worker" is provided enhanced guidance for critical actions. New technologies highlighted in this presentation include autonomous drones, cyber monitoring, turn around management and safety optimization. A digital thread ties all the data streams into one strategic view to drive efficient operations, faster anomaly detection and better data integrity using advanced analytics. Emergency management has also been dramatically improved with real-time optimization of worker location and activity providing a more flexible safe environment with responsive communication to manage workers and guide them safely during a process safety event.

Delek will discuss their vision and will demonstrate the platform with the help of GOARC, a technology partner who is helping develop their concept for the Refinery of the Future.

620058 Investigation of Transfer Learning for Image Classification and Impact on Training Sample Size for Image Classification Tasks

Abstract:

Recent developments in deep learning have brought huge breakthroughs in the image processing area, which triggered numerous successful applications and positively impacted the current big data context of Industry 4.0. On the other hand, it is widely known that large amounts of training data are required to train a deep learning model with millions of parameters from scratch, which limits its application in many industrial applications where sufficient labeled data are often lacking. Transfer learning is one of the practical solutions to reduce the data required for training, which tries to reuse learned knowledge for similar tasks. Nevertheless, many technical details of transfer learning implementation are not well documented.

In this talk, different transfer learning approaches and implementation details for high-performance model building under the constraint of limited available training data are investigated using two different industrial use-cases. Various methods of transfer learning are compared and important technical details are discussed. Through this study, the minimum number of training samples can be estimated and practical guidelines for the development of image classification models with limited data resources are summarized.

IMPLEMENTACIÓN Y USOS DE SIMULADOR DINÁMICO DE PROCESOS EN UNIDAD DE METANOL PLAZA HUINCUL

Rubén Edgardo Basso
Supervisor Producción
ruben.basso@ypf.com

Guillermo Fabian Verni
Supervisor Producción
guillermo.verni@ypf.com

Eduardo Cura
Jefe de Unidad Refinería
eduardo.r.cura@ypf.com

Eliana Peralta
Jefe de Planta de Agua Río Neuquén
eliana.peralta@ypf.com

Luis Alfredo Saavedra
Gerente Producción
luis.saavedra@ypf.com

YPF, Complejo industrial Plaza Huincul (CIPH), Neuquén, Argentina

Prepared for Presentation at
American Institute of Chemical Engineers
2021 Spring Meeting and 17th Global Congress on Process Safety
Virtual
April 18 - 22, 2021

IMPLEMENTACIÓN Y USOS DE SIMULADOR DINÁMICO DE PROCESOS EN UNIDAD DE METANOL PLAZA HUINCUL

Rubén Edgardo Basso
Supervisor Producción
ruben.basso@ypf.com

Guillermo Fabian Verni
Supervisor Producción
guillermo.verni@ypf.com

Eduardo Cura
Jefe de Unidad Refinería

Eliana Peralta
Jefe de Planta de Agua Río Neuquén

Luis Alfredo Saavedra
Gerente Producción

YPF, Complejo industrial Plaza Huincul (CIPH), Neuquén, Argentina

Keywords: simulador – entrenamiento – beneficios – implementación – habilidades – mejoras.

Abstract

El propósito de este artículo es comprender el impacto de la incorporación de un simulador dinámico de entrenamiento operativo en un proceso productivo. Esta decisión puede reducir la demanda de los recursos necesarios para alcanzar la formación del operador y los riesgos asociados al entrenamiento en el puesto de trabajo, incrementando las oportunidades de actuación y práctica.

Cada proceso mantiene una complejidad particular para su aprendizaje, situación que debe superarse durante la instrucción para alcanzar los objetivos que hayan sido establecidos. El modelo adoptado y los planes de formación deben diseñarse a medida.

Todo el conocimiento disponible sobre el proceso productivo enriquece el simulador, originando nuevos y mejores escenarios, para obtener de esta manera un robusto plan de formación.

Los operadores inician su entrenamiento y en contados minutos quedan totalmente inmersos en una simulación con diferentes escenarios a los que podrían enfrentarse durante la rutina diaria. Desde su puesto de trabajo, sabrán como actuar, no solo por la lectura de un procedimiento, su interpretación o comentarios del instructor sino en base a las prácticas realizadas en el simulador dinámico.

Los resultados obtenidos permiten identificar diversos beneficios, tales como facilitar la instrucción operativa mejorando la efectividad de la formación de los operadores, el desarrollo de habilidades promoviendo un proceso productivo más seguro, con actuaciones más precisas y la reducción de errores operativos.

Introducción

El Simulador de Entrenamiento Operativo (OTS) es una herramienta que aumenta la capacidad de compartir conocimientos, habilidades y experiencias dentro de una organización.

La gestión del conocimiento fundamentalmente se basa en poder transferir el conocimiento o experiencias de una persona o de un grupo de personas a otras, lo cual no es fácil de realizar. La información puede almacenarse, modificarse y transmitirse fácilmente, pero las habilidades, en cambio, son complicadas de obtener y llevan más tiempo, demandando más horas de práctica.

Planificar la formación requiere de esfuerzo para identificar las personas que poseen habilidades valiosas, dentro y fuera de la organización, para que compartan sus conocimientos; siendo éste, el único activo que crece con el tiempo, pero desaparece con las personas si no es efectivamente compartido.

El OTS es un dispositivo de entrenamiento integral diseñado para proporcionar una experiencia práctica basado en modelos de simulación dinámica de la unidad de producción. Este dispositivo combina software y hardware para representar el proceso y el sistema de control de manera dinámica.

Figura **1**. Sala de Control Unidad Metanol.

El software proporciona capacitación a los operadores del Sistema de Control Distribuido (DCS) sobre diversos aspectos de la operación de la planta, incluyendo fallas específicas que permiten al instructor del OTS administrar diferentes escenarios de operación. Esto se logra mediante la creación de condiciones de operación y con la activación de mal funcionamientos preconfigurados en la estación de simulación. El operador de consola, a través de las estaciones de operación procede a la puesta en marcha, parada de emergencia, operación normal, etc.; aprendiendo a reconocer alarmas, tomar acciones y resolver distintos problemas operativos que normalmente puede presentarse en la unidad.

En el presente documento, se darán detalles de la experiencia, lecciones aprendidas y mejores prácticas en la implementación del simulador por los instructores y los operadores de consola.

1 Antecedentes

El método de aprendizaje del operador de consola se fue adaptando a la tecnología disponible y a las nuevas generaciones incorporadas a la empresa.

Originalmente, un operador de consola tenía su programa de formación basado en los procedimientos teóricos y la experiencia de otros compañeros, transmitido de boca en boca, y solapando sus tareas a la de un tutor (también operador de consola).

Un ejemplo de los procedimientos a estudiar está representado en la Figura 2.

Memorizar las acciones que se deben ejecutar o que el procedimiento refleje fehacientemente lo que se realiza en planta es prácticamente imposible. A su vez, poner en práctica el entrenamiento con una planta estable tampoco ayuda a afianzar los conocimientos por su imposibilidad de realizar acciones. Con las nuevas tecnologías disponibles, las prácticas se vuelven realistas, incluyendo en algunos casos desarrollos audiovisuales, haciendo más efectivo el proceso de aprendizaje.

Figura 2. Instructivos de puesta en marcha.

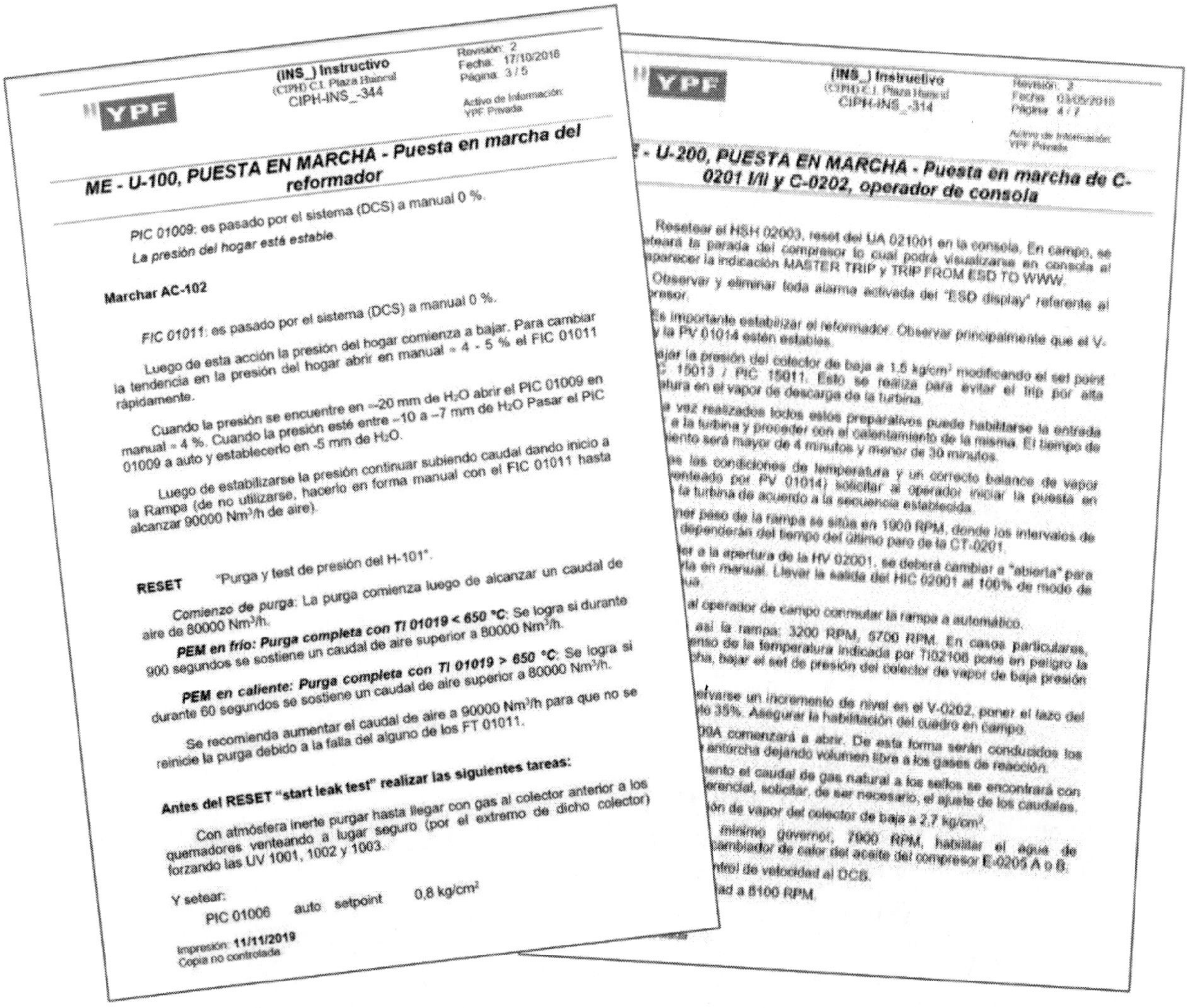

(INS_) Instructivo
(CIPH) C.I. Plaza Huincul
CIPH-INS_-344
Revisión: 2
Fecha: 17/10/2018
Página: 3 / 5
Activo de Información: YPF Privada

ME - U-100, PUESTA EN MARCHA - Puesta en marcha del reformador

PIC 01009: es pasado por el sistema (DCS) a manual 0 %.

La presión del hogar está estable.

Marchar AC-102

FIC 01011: es pasado por el sistema (DCS) a manual 0 %.

Luego de esta acción la presión del hogar comienza a bajar. Para cambiar la tendencia en la presión del hogar abrir en manual = 4 - 5 % el FIC 01011 rápidamente.

Cuando la presión se encuentre en –20 mm de H_2O abrir el PIC 01009 en manual = 4 %. Cuando la presión esté entre –10 a –7 mm de H_2O Pasar el PIC 01009 a auto y establecerlo en -5 mm de H_2O.

Luego de estabilizarse la presión continuar subiendo caudal dando inicio a la Rampa (de no utilizarse, hacerlo en forma manual con el FIC 01011 hasta alcanzar 90000 Nm³/h de aire).

RESET "Purga y test de presión del H-101".

Comienzo de purga: La purga comienza luego de alcanzar un caudal de aire de 80000 Nm³/h.

PEM en frío: Purga completa con TI 01019 < 650 °C: Se logra si durante 900 segundos se sostiene un caudal de aire superior a 80000 Nm³/h.

PEM en caliente: Purga completa con TI 01019 > 650 °C: Se logra si durante 60 segundos se sostiene un caudal de aire superior a 80000 Nm³/h.

Se recomienda aumentar el caudal de aire a 90000 Nm³/h para que no se reinicie la purga debido a la falla del alguno de los FT 01011.

Antes del RESET "start leak test" realizar las siguientes tareas:

Con atmósfera inerte purgar hasta llegar con gas al colector anterior a los quemadores venteando a lugar seguro (por el extremo de dicho colector) forzando las UV 1001, 1002 y 1003.

Y setear:
PIC 01006 auto setpoint 0,8 kg/cm²

Impresión: **11/11/2019**
Copia no controlada

(INS_) Instructivo
(CIPH) C.I. Plaza Huincul
CIPH-INS_-314
Revisión: 2
Activo de Información: YPF Privada

- U-200, PUESTA EN MARCHA - Puesta en marcha de C-0201 I/II y C-0202, operador de consola

El modelo desarrollado por Morgan McCall, Robert Eichinger y Michel Lombardo, de la Universidad de Princeton, considera que la forma más efectiva de aprender y desarrollar nuevos comportamientos y habilidades es mediante la práctica en situaciones laborales.

En este sentido, como el OTS recrea las condiciones del puesto de trabajo, los planes de formación fueron actualizados para extender las prácticas en el simulador.

Figura 3. Modelo 70:20:10

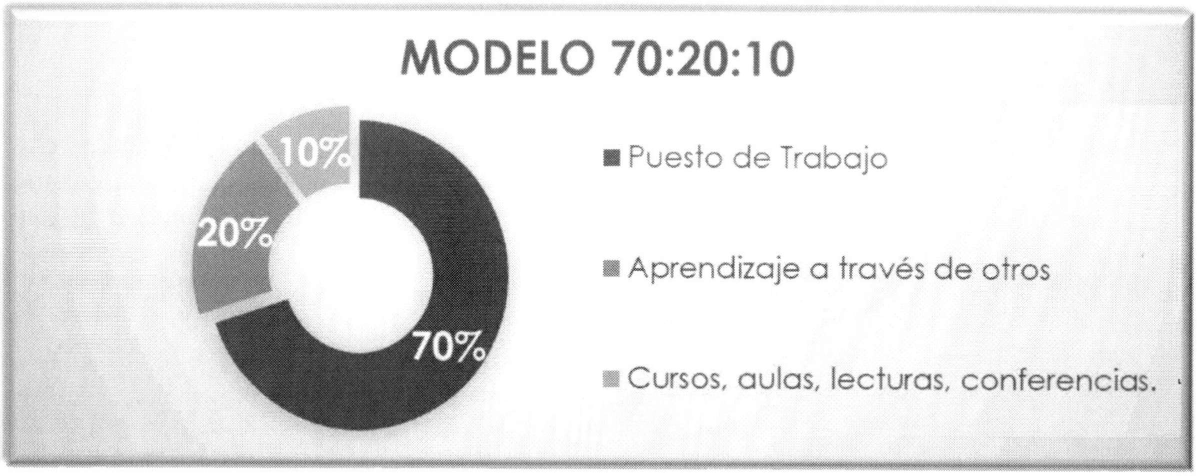

El aprendizaje mediante la realización de maniobras asistidas en campo, los simulacros operativos, las experiencias vividas y las prácticas en simulador alcanzan a cubrir el 70% de efectividad.

2 Desarrollo

Es sabido que el uso de simuladores en la industria no es nuevo, en especial si se toma en cuenta la aeronáutica. Cada compañía, ha desarrollado sus mejores prácticas en el uso de simuladores en base a sus experiencias y necesidades de formación.

El simulador dinámico para la planta de Metanol en el Complejo Industrial Plaza Huincul (CIPH) tiene su inicio del proyecto a mediados del 2013 finalizando con su puesta en servicio en junio de 2015.

Para afrontar el desafío se conformó un equipo para el proyecto y puesta en servicio del OTS que incluyó a personal de los departamentos de Optimización y Control, Ingeniería de Procesos y Producción.

Las tareas realizadas por el equipo para lograr la puesta en funcionamiento del OTS fueron:

- Definición de la especificación funcional del diseño.
- Prueba de aceptación del modelo (MAT)
- Prueba de aceptación en fábrica (FAT)
- Prueba de aceptación en sitio (SAT)
- Simulación dinámica.

Definición de la especificación funcional del diseño

La estación del instructor está conformada por una Interfaz Hombre Máquina (HMI) propio del software del simulador, respetando los P&IDs de la unidad y su nomenclatura. Ver arquitectura de un OTS en figura 4.

En dicha estación se puede:

- Activar la función que desee para entrenar al operador.
- Correr o detener la simulación.
- Pausar la simulación, pudiéndose visualizar y capturar los datos del momento.
- Activar malfuncionamientos. Pueden presentarse por separado, simultáneamente o programarse secuencialmente en cualquier momento específico o con un intervalo de tiempo definido.
- Modificar parámetros externos y variables controladas (temperatura ambiente, temperatura del agua de enfriamiento, ciertas condiciones de la alimentación, etc.)
- Operar todos aquellos equipos que el operador de consola no puede controlar de forma remota, por ejemplo: la apertura de una válvula manual, el encendido de quemadores, la marcha de motores, etc.
- Visualizar tendencias de las distintas variables en intervalos de tiempo específicos.
- Realizar el seguimiento de las variables operativas activando la función de supervisión para evaluar el rendimiento del operador, reportando el número de veces y tiempo que la variable excedió los límites permitidos.

Figura 4. Arquitectura OTS.

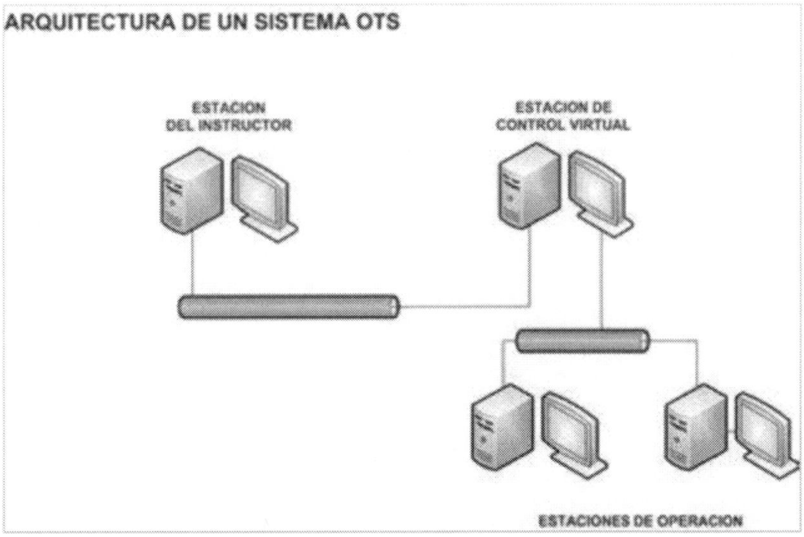

Prueba de aceptación del modelo (MAT)

Luego de establecida la especificación del diseño del Simulador se procede a la validación del modelo realizado en las instalaciones del CIPH. Se ratificó el alcance del modelo del proceso, se evaluó y probó el modelo en estado estacionario y el comportamiento dinámico de los principales elementos del proceso. La aceptación de las pruebas MAT es necesaria para que posteriormente los modelos de proceso puedan ser integrados con el sistema de control DCS.

Prueba de aceptación en fábrica (FAT)

En esta fase de aceptación se revisó el sistema de entrenamiento completo que consta del modelo de procesos completamente integrado con el sistema de control y el sistema de seguridad, todo corriendo en el hardware del proyecto. Esto incluye una inspección del hardware y software definido en el alcance del proyecto.

Prueba de aceptación en sitio (SAT)

Una vez superada la etapa de Aceptación en Fábrica, el simulador pasó a la etapa de montaje, disposición y verificación en la unidad de negocio. El objetivo de esta etapa fue demostrar la funcionalidad del simulador como un sistema completo luego de instalarse en el sitio.

Simulación dinámica

Una simulación dinámica fidedigna representa de la mejor manera la hidráulica, la transferencia de calor y otras condiciones de operaciones reales de la planta, habilitándola a ser utilizada como una herramienta de análisis para Ingenieros de Procesos, Ingenieros de Control e Ingenieros de Operación.

Como beneficio sobre la simulación en estado estacionario, en la simulación dinámica se puede evaluar el comportamiento en todo rango operativo y no operativo o de emergencia. Se pueden plantear distintos escenarios para probar nuevas estrategias de control, modificaciones en variables operativas como presiones, composiciones, equipos, etc., generando oportunidades de mejora u obteniendo soluciones para la toma de decisiones por problemas operativos poco usuales o no en la unidad de proceso.

3 Implementación del OTS en la Unidad de Metanol

La implementación del simulador en la planta de metanol se realizó mediante consenso entre la jefatura de planta y los instructores, para poder hallar el mejor modo de llevar adelante la formación, resolviendo proyectarla en un plan plurianual. Para esto, se definieron tres grandes bases:

- Formación de instructores.
- Armado del plan de formación.
- Temario de simulación a desarrollar.

Formación de instructores

La primera consideración que se tuvo presente fue la formación de instructores. Sin lo cual, no era posible dar autonomía al programa.

Una vez ejecutada la SAT y FAT, se determinó que las personas que participaron en las pruebas de aceptación junto a los representantes de la contratista del simulador se ocupen de la capacitación de los instructores.

Se acordó que los supervisores de turno asuman el rol de instructor, con un tiempo estimado de 6 meses de formación. Designar como Instructor a un supervisor de turno da autonomía para la organización del plan de formación y permite continuar con lo visto en la práctica durante las tareas de operación normal.

Cada supervisor de turno es responsable de la formación de sus tableristas y a su vez, cada supervisor es instructor, esto nos permite tener flexibilidad para establecer el cronograma.

Armado del plan de formación

La Dirección del CIPH y el departamento de RRHH permitió desarrollar un plan de formación anual para los tableristas disponiendo de sus descansos. De esta manera, el supervisor, teniendo las bases de formación prepara el diagrama anual de sus operadores de consola.

Temario de simulación a desarrollar

Para poder entender el temario acordado es necesario explicar en breves palabras cómo es la operación de la planta de metanol.

La puesta en marcha de la unidad es uno de los temas más tratados dada la complejidad de la operación en esta etapa.

La Planta de Metanol, desde su puesta en marcha en frío hasta obtener el metanol en especificación, tiene un tiempo promedio de 24 horas.

La secuencia de puesta en marcha se puede resumir en 3 grandes pasos:

i. Puesta en marcha del reformador: en esta etapa se debe contemplar la secuencia de encendido del horno, la curva de calentamiento a 50°C/h y el ingreso de vapor y gas.
ii. Puesta en marcha del compresor: es crítico conocer el funcionamiento de los sellos, el cambio de presiones, cambio de gas combustible, permisivos de marcha y paros de la turbina y la habilitación del vapor de extracción.
iii. Acondicionamiento del metanol: una vez llegado el metanol a la unidad de destilación, es muy importante encontrar los perfiles de destilación, habilitar los gases al sistema de gas combustible y lograr la especificación.

Por su orden y tiempos, rara vez (prácticamente nunca), un operador de consola puede participar de la secuencia completa de puesta en marcha.

El simulador dispone de una serie de facilidades que permiten controlar la velocidad del tiempo de simulación, reduciendo la espera entre operaciones y en el caso de ser necesario, enlentecerlo para lograr una mejor observación del proceso. Además, cuenta con la posibilidad de grabar las sesiones de cada operador en el momento deseado y continuar en el próximo día de entrenamiento desde el lugar donde se dejó u otro estado previamente guardado. Estas ventajas que ofrece permiten ver y participar de todos los pasos de puesta en marcha de todas las unidades. En otras palabras, le permite al operador ver la película completa y no solo fragmentos.

Este tipo de simulador es muy requerido por plantas que trabajan con secuencias de puesta en marcha diferidas entre varios turnos.

Los escenarios se desarrollan sobre una plantilla, requiriendo el orden secuencial de acciones a dos columnas con celdas de verificación; coordinando las tareas de campo con las del operador de consola. Los escenarios que componen la simulación de puesta en marcha son:

1. Preparación de calderas.
2. Preparación sistema de vapor.
3. Preparación U-300, destilación.
4. Puesta en marcha AC-0101 / AC-0102, aire de combustión.
5. Encendido reformador.
6. Ingreso de vapor a E-0102, protección de equipos por altas temperaturas.
7. Ingreso de vapor H-0101, horno reformador.
8. Ingreso de GN a reformador.
9. Administración de los colectores de vapor.
10. Ingreso de gas reformado a quemadores.
11. Calentamiento torres destilación.
12. Preparación reactor de síntesis y turbina.
13. Preparación del compresor.
14. Puesta en marcha del compresor.
15. Puesta en marcha destilación y U-500 (planta de agua desmineralizada).
16. Puesta en marcha luego de vaciado de calderas.

Otros escenarios que forman parte de las prácticas de simulación.

- Escenarios de paro de la Unidad.

17. Parada controlada de planta.
18. Parada de planta, primera respuesta a la emergencia operativa.
19. Parada de planta, corte del suministro eléctrico.
20. Paro de caldera auxiliar, acciones de emergencia sobre el proceso.
21. Paro caldera auxiliar a mínima carga.
22. Paro U-200, síntesis de metanol.

- Escenarios de Emergencias operativas como ser:

23. Falla en el suministro de energía eléctrica.
24. Falla en el sistema de agua enfriamiento.
25. Falla en el suministro de aire de instrumentos.
26. Falla en el sistema de control (DCS).
27. Falla en el sistema de vapor.

- Interlocks, prácticas y chequeos de funcionamiento en tiempo real, de acuerdo con lo establecido en el sistema de emergencia independiente de paro de planta e interpretación de lazos complejos. Para estos escenarios se utiliza documentación de O&C.

28. Interpretación acciones lazo complejo sistema de vapor
29. Interpretación acciones lazo complejo relación V/C y carga.
30. Interpretación acciones lazo complejo gas combustible.
31. Interpretación de interlocks.

- Mal funcionamiento de válvulas e indicaciones, escenarios forzado por el instructor, con consecuencias en la operación de consola. Pantalla dedicada en la estación del instructor.

- Análisis de sensibilidad de los procesos: dado que el sistema es una réplica de las condiciones de diseño de la unidad, se puede ver cómo responde la planta ante un cambio de composición de la materia prima, o variaciones en la presión de suministro de gas natural.

4 Usos del simulador

- Entrenamiento inicial para nuevos operadores. El plan de formación se basa en la práctica de los escenarios completos de puesta en marcha en frío, parada de planta, puesta en marcha inmediata de las unidades bajo su responsabilidad.

Figura 5. Ficha del operador.

OTS: Ficha del Operador	
Fecha sesión de entrenamiento	/ /
Operador:	LP:
Instructor:	LP:
Escenario de entrenamiento[1]:	

Acciones sobre el marco de Aplicación		**SI**	**NO**
Este apartado aplica sólo para Escenarios de entrenamiento que impliquen Puesta en Marcha, Operación normal, Parada de Emergencia y Parada de planta controlada.	El Operador reconoce el/los procedimiento/s que se aplican para este escenario.[2]		
	Las acciones que ejecuta el Operador sobre el Sistema de Control son correctas.[3]		
	Las prioridades de las acciones llevadas adelante por el Operador son correctas.[4]		
Este apartado sólo aplica para los escenarios de Entrenamiento donde se inducen fallas o mal funciones.	Como la mal función se introduce en cualquier momento mientras la operación de la planta se encuentra en régimen estable, en ese momento, el Operador detecta alguna evidencia.[5]		
	Reconoce de forma inmediata la causa del inicio de la falla, la cual fue inducida por el Instructor.[6]		
	Las acciones adoptadas por el Operador para corregir el desvío y normalizar la Unidad son correctas.[7]		
Existe una comunicación permanente con el Operador de la otra estación de Control.[8]			
Mantiene comunicación fluida y solicita acciones oportunas al Instructor (Operador de Campo).[9]			

Nivel de Conocimientos sobre el escenario de Entrenamiento[10]	**Entiende**	**Aplica**	**Domina**

Observaciones[11]

- Entrenamiento periódico de operadores de consola con experiencia. A principio de año, cada instructor recibe las directrices de formación para su grupo. Luego se establecen los días mensuales extras a la jornada laboral. No obstante, en el transcurso de la formación cada instructor tiene la potestad de fortalecer o cambiar los escenarios.

  Los programas para realizar de cada operador se basan en la secuencia de puesta en marcha. El objetivo principal es que el operador aprenda toda la puesta en marcha de la unidad. Si tiene algún error, este se corrige y se repite la secuencia. El instructor es quién lleva el avance y conocimiento de cada uno de sus colaboradores.

- Seguimiento del programa de formación. El seguimiento de los operadores se basa en el cumplimiento de los escenarios establecidos, y se realiza de manera personalizada. Ver el registro ficha del operador en Figura 5.
- Para la evaluación del personal en escenarios de operación normal y emergencia se implementaron formularios para poder verificar la efectividad de la formación recibida. Ver plantilla de escenarios y evaluación en Figura 6.

Figura 6. Plantilla de escenarios y evaluación.

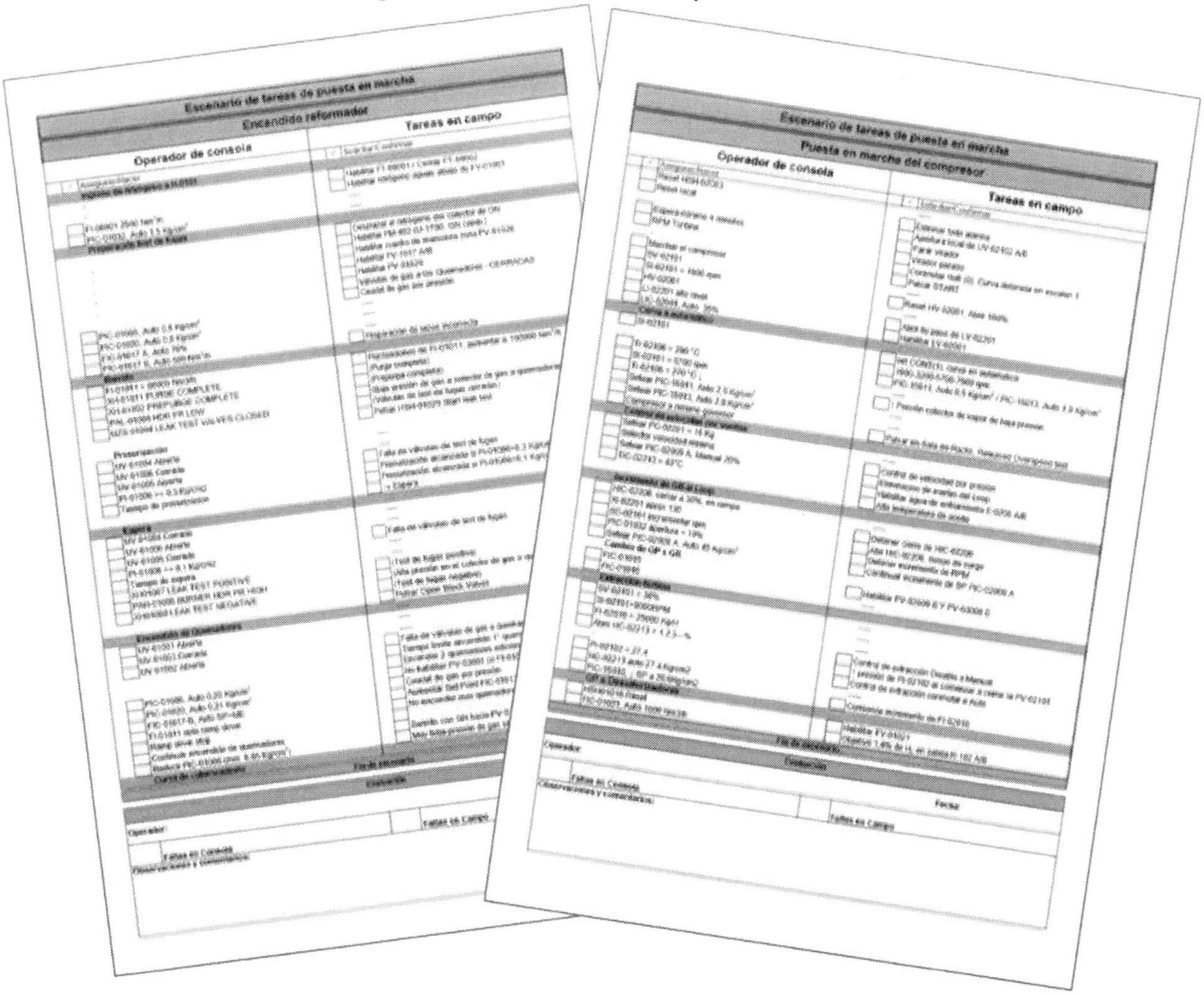
Escenario de tareas de puesta en marcha
Encendido reformador
Operador de consola
Tareas en campo

Escenario de tareas de puesta en marcha
Puesta en marcha del compresor
Operador de consola
Tareas en campo

- Gestión del plan de sucesiones para operadores de consola. El concurso por examen para Operadores de Campo que aspiran a Operador de Consola se realiza en dos instancias. Examen teórico y, por último, los primeros del ranking rinden un examen en el Simulador Dinámico. Ver Figura 7. Evaluación en simulador dinámico.

  Dentro del plan de sucesiones, definido el número requerido de operadores en formación (depende de las necesidades futuras), se designa un instructor para el grupo y se establece el plan de formación anual. Ver ejemplo de formulario de evaluación en Figura 8.

- Desarrollo de nuevos procedimientos operativos. Para ello, se trabaja en conjunto con el departamento de Ingeniería de Procesos y con el tecnólogo para validar la práctica. Luego se realiza la gestión del cambio correspondiente.

  Dentro de una emergencia difícilmente se puedan probar opciones de mejora. Por ejemplo, durante las prácticas de emergencia de paro del reformador por el corte del suministro de energía eléctrica, se ensayaron diferentes acciones del operador de consola, para evitar el vaciado brusco de la caldera de alta presión. Definido el procedimiento, se llevó a cabo en la práctica logrando su propósito, reduciendo el riesgo de vaciado de caldera de alta presión y mejorando los tiempos de puesta en marcha en más de 10 horas.

Figura 7. Evaluación de conocimientos para aspirantes a Operador de Consola.

YPF	*EC – Evaluación de Conocimientos*	Fecha: 25/11/2020
	EC - CONSOLA	Página 1 de 13

Operador de planta: ....

Legajo personal: ....

1. En la corriente de salida de la planta 2 se determina la concentración de cloruros. Indicar:
   a. ¿Cuál es el sentido de dicho análisis?
   b. ¿Cuál es la fuente de ingreso de cloruros al sistema?
   c. ¿Una alta concentración de este producto en el sistema, trae aparejado algún problema?
2. Realizar una descripción de todas las corrientes que puedan ingresar a TK-1601 y de las que pueden derivarse sin ingresar. ¿A cuál de las corrientes debo enfriar? ¿Cuál es el posible origen del agua agregada para enfriamiento?, mencionar ventajas y desventajas de las corrientes de enfriamiento.
3. Mencionar en orden las etapas de regeneración de los lechos catiónicos y aniónicos de planta 2. ¿Cuál es la importancia de respetar el orden?
4. ¿Cuáles son la etapas de regeneración del lecho mixto? Mencionar los cuidados a tener en cada etapa si corresponde.
5. ¿Por qué luego de una regeneración de la planta 1 se necesita adicionar soda para neutralizar la pileta? Con el mismo criterio se necesita incorporar ácido al efluente de la planta 2. ¿Por qué? Justificar. Si lo antes mencionado no se cumple, ¿A qué puede deberse?
6. La unidad de destilación puede ser parada ya sea para mantenimiento o debido a una parda general de la planta.
   a. ¿Cuáles son la etapas operativas para desarrollar en caso de sacar de servicio la unidad de destilación para una parada programada suponiendo que la unidad se encuentra produciendo 1150 Tn/d de Metanol?
   b. Con la unidad de destilación fuera de servicio al igual que la unidad de síntesis se observa incremento en el nivel de V-0301, ¿A qué se debe esta condición? ¿Qué tareas debo realizar?
7. Supongamos que se presenta una situación de corte de energía eléctrica, pero en el cual se evitó el vaciado de las calderas y la reposición de energía se produce 4 horas después del incidente del corte, por lo que se decide un arranque en caliente. ¿Qué acciones instantáneas debería realizar en la unidad de destilación luego del corte de energía? Indicar acciones desde sala y campo, listarlas por orden de prioridad.
8. ¿Cuál es la consecuencia de salir al Cañadón Zapala con el efluente fuera de especificación? Nombre algunos componentes y parámetros máximos. Mencione al menos 5 parámetros con su máximo valor admisible. ¿Cuál es el caudal máximo permitido?¿Existe alguna limitación en la medición?
9. Durante la puesta en marcha se prepara a llenar las torres de destilación, recibiendo desde TK-0401 metanol crudo. Mencione como debe permanecer el cuadro de la PV-03001 durante toda la etapa de encendido de quemadores del reformador. ¿Por qué?
10. Situación: suministro normal del agua de EPAS, calidad 250 µs/cm; se produce la contaminación del TK-0501, 11 µs/cm, pH 3, nivel 92%, ósmosis inversa en marcha produciendo 30 m3/h, 2,5 µs/cm; condensado de proceso a TK-1601 por alto tiempo de filtrado.
    a. ¿Qué efectos se producirán aguas abajo del proceso?

EXAMEN PARA ACCEDER A OPERADOR DE CONSOLA COMPLEJA DE METANOL

- Creación de nuevos escenarios de práctica a partir de eventos reales y lecciones aprendidas. No todos los operadores viven los eventos de una determinada emergencia operativa. Una vez cerrada la investigación del incidente, se procede a la divulgación y si merece, se gestiona un nuevo escenario de práctica. Algunos ejemplos son:

  Ejemplo 1

  Un turno de operadores enfrentó una emergencia operativa debido a la falla de una celda eléctrica de 380V, afectando el estado de marcha de motores de la unidad de destilación;

provocando el cese del caudal de reflujo, alimentación, salida y problemas de condensación.

Figura 8. Evaluación en simulador dinámico.

Parte 2 – Evaluación en simulador dinámico.

Examen para Operador de Planta para acceder a Operador de Consola Compleja.

Alcance

1) Reconocimiento de planta.
2) A partir de la condición inicial "ESTADO ESTACIONARIO" se genera el paro de la bomba P-0304.
3) Escenario alta presión de V-0301. Se genera una alta presión el equipo mencionado.
4) Paro de Emergencia de la Unidad de Destilación (U-0300). A partir de la condición inicial "ESTADO ESTACIONARIO" se genera un paro de la unidad (generando una malfunción de la PV-01009).

1) ¿Conoce la ubicación de los equipos y válvulas de control en las pantallas y en el campo?

	Pantalla	SI	NO	Ubicación en campo	SI	NO
LV03008 – reflujo de T-0301	2			1° plataforma, sur de V-0302		
P-0301 – fondo de T-0302	1					
E-0302 – enfriador de cabeza de T-0301	2			2° plataforma		
E-0307 – enfriador de cabeza de T-0302 (AA E-0306)	4			1° plataforma		
FV03016 – reflujo de T-0302	3 y 4			1° plataforma, sur de E-0307		
E-0306 - enfriadores de cabeza de T-0302	4					
J-0301 – eyector gases no condensables de T-0301	2			3° plataforma		
V-0301 – carga a T-0301	1					
P-0304 – fondo de T-0301 a T-0302	2					
E-0305 – reboiler de vapor de T-0302	3					
PV03001 – gases de expansión de V-0301	1			1° plataforma, norte de E-0307		
P-0303 – metanol puro salida DE v-0303	4					

2) Paro de P-0304

Acciones	SI	NO
Dar aviso al operador de la unidad		
Cerrar la LV03003 luego de conmutar el controlador a manual.		
Solicitar al operador la puesta en servicio de la P-0304 B.		
Abrir la LV03003 al valor de operación normal y conmutar el controlador a AUTO.		
Comunicación con operador de planta.		
¿Es preciso con las tareas solicitadas?		

La columna de metanol puro comenzó a vaciarse, sabiendo que su nivel de fondo intercambia calor con la corriente de gas reformado, proceso que continuaba en marcha. La unidad de proceso se recupera de la emergencia marchando los equipos de reserva.

Otro turno de operadores realizó el análisis y simulación, donde se pudo constatar que en los minutos que duró la falla eléctrica, el equipo de intercambio de calor "gas reformado - fondo de torre de destilación" no se vio afectado a pesar de perder la indicación de nivel. Sin embargo, en la simulación se verifica que algunos minutos más hubieran alcanzado para el vaciado total de la columna, perjudicando el intercambio de calor mencionado y en consecuencia el proceso de la unidad de reformación. Se generaron consultas al departamento de Ingeniería de Procesos para validar la decisión de paro de la unidad.

Ejemplo 2

Se realizó el paro de emergencia controlado de la unidad de proceso debido a la rotura de un tubo en el reformador. Durante la emergencia, las acciones operativas hicieron foco en el control del reformador debido a que la rotura del tubo provocó fluctuaciones en el proceso. El trabajo ejecutado durante la emergencia operativa está en proceso de evaluación para poder desarrollar un nuevo escenario de mejores prácticas en el OTS.

Ejemplo 3

Falla de alimentación de agua a calderas. Se establecen las acciones de contingencia que deben realizarse en el proceso frente al paro intempestivo de la bomba de alimentación de agua a las calderas en ausencia de la bomba de reserva y a pesar de realizar la puesta en marcha inmediata del equipo detenido. Las acciones son ensayadas y validadas en el simulador para poder establecer un nuevo escenario de práctica en el OTS.

Ejemplo 4

Durante el estado de mínima carga al reformador, el sistema de vapor requiere la máxima producción de la caldera auxiliar. En estas condiciones, se produjo el paro de la caldera entrando en situación de emergencia todo el sistema de vapor. Las acciones ejecutadas lograron estabilizar el proceso, lo que evitó el paro de la unidad. El procedimiento fue validado en las simulaciones y se estableció como un nuevo escenario del entrenamiento del OTS.

Ejemplo 5

Una falla en la válvula de alimentación de la columna de destilación primaria ocasionó la emergencia operativa de la unidad de proceso. El personal operativo debió determinar la causa y las acciones que permitan demorar o atenuar las consecuencias para poder solucionar el problema. Las acciones son validadas en el simulador y se inicia la preparación del escenario para formar parte de las capacitaciones en OTS.

- Determinación de la efectividad de los procedimientos propuestos. Desarrollo de un instructivo unificado para las tareas de puesta en marcha y parada planificada de la unidad, donde el requisito mínimo de aceptación es medir su eficacia en el simulador. Modernización gradual de la plantilla original de instructivos, por una paso a paso, eliminando extensos párrafos y que integra las acciones de todos los responsables de la ejecución del procedimiento. Ver ejemplo de plantillas de procedimientos en Figura 9.
- Confirmación de la efectividad de nuevos controles propuestos luego de realizar los ensayos en el OTS.
- Interpretación de iniciadores y acciones de interlocks, y el funcionamiento de los lazos complejos. Capacitación sobre el sistema de seguridad del proceso y lazos complejos empleando las herramientas disponibles del simulador. Ralentizar el proceso y repetirlo las veces que sea necesario permite al operador en entrenamiento visualizar y entender las acciones de los lazos complejos.
- Estudio de limitaciones en el proceso. Analizar la respuesta del proceso y ensayar posibles soluciones operativas frente a variaciones de: presión del suministro de materias primas, temperatura de trabajo en el reformador, temperatura ambiente, etc.
- La medición de la eficacia de la capacitación operativa es uno de los desafíos más importantes de la gestión del conocimiento. Cuanto del contenido desarrollado durante la

actividad de formación es incorporado al conocimiento del operador. El simulador permite exponerlo de manera inmediata, dejando en claro al momento de la instrucción, los puntos fuertes y a mejorar del conocimiento de la persona.

Figura 9. Ejemplo plantilla original vs plantilla objetivo.

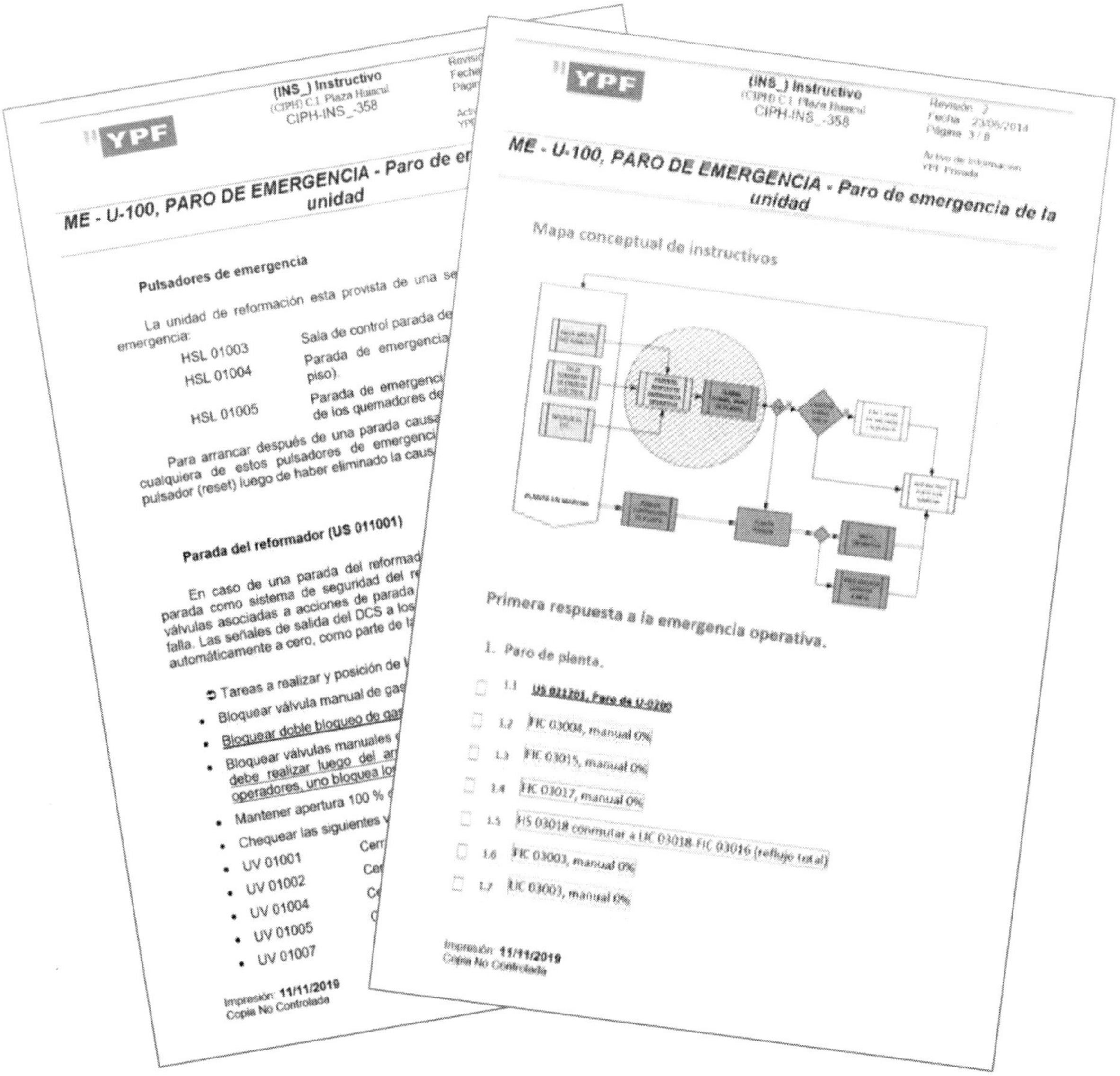

5 Historial de formación

Durante los períodos 2016 y 2017, fueron capacitados 14 operadores de consola (nuevos y experimentados) sumando un total de 710 horas de capacitación. (350 HH por año aprox.)

Para el período 2018 se lanzó un plan de capacitación cumpliendo aproximadamente 460 horas de entrenamiento.

Durante 2019, se cumplieron un total 576 horas para operadores de consola, además de 288 horas de instructores no incluidas en el gráfico.

Condicionados por la pandemia del COVID-19, se debió adecuar el lugar del simulador, trasladando a otra oficina el servidor de la estación del instructor, a fin de cumplir con los protocolos de distancia social mínima para evitar contagios . En 2020 se realizaron 122 horas de formación en el simulador.

Figura 10. Evolución en la utilización del simulador dinámico.

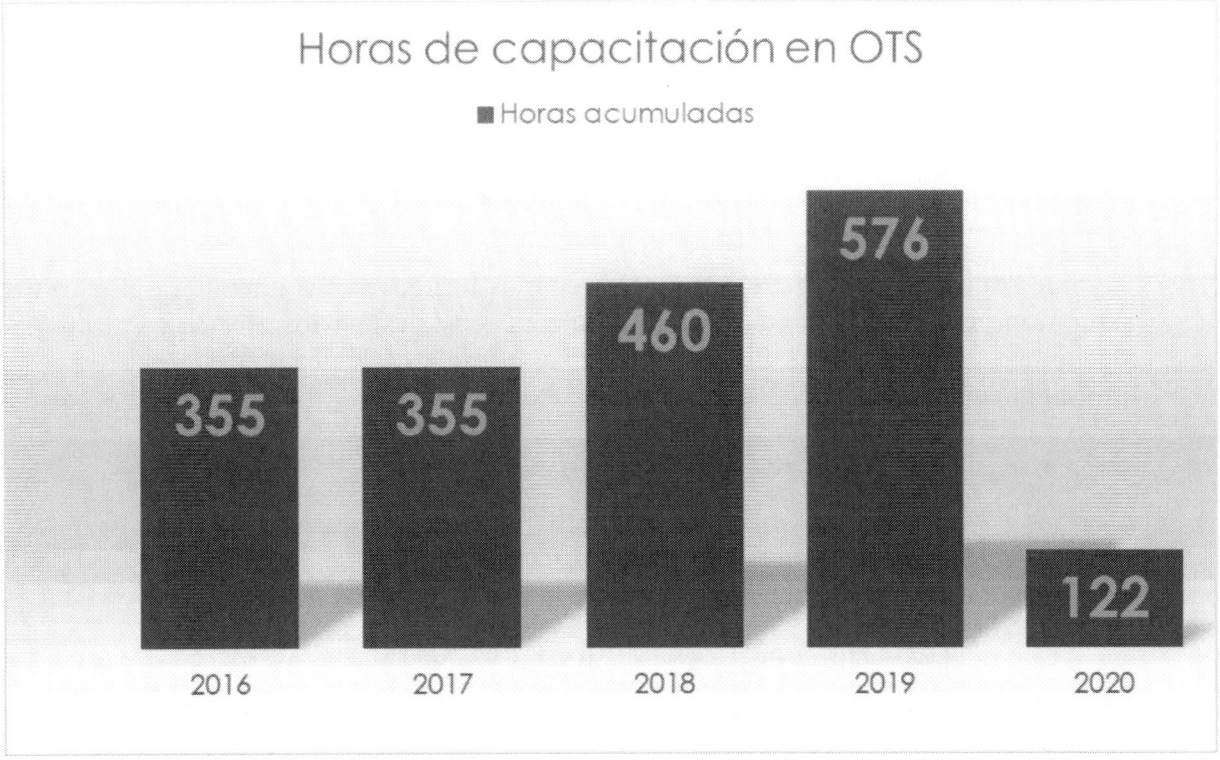

6 Conclusiones

- La implementación del OTS permitió reducir errores operativos durante los arranques y paros de planta programados. El operador puede practicar en el OTS aquellos fragmentos del procedimiento de puesta en marcha o paro de la unidad de proceso en los que menos experiencia haya adquirido debido a la falta de oportunidades en el proceso real o por haber transcurrido mucho tiempo desde la última vez que realizó la tarea operativa.
- Los arranques son más eficientes debido a la corrección de controles. A medida que el operador de consola adquiere experiencia puede desarrollar nuevas formas de ejecutar una tarea, establecer nuevos parámetros o el reconocimiento de nuevas variables. El OTS permite ensayar y evaluar nuevas prácticas conformes a obtener una mayor eficiencia operativa. Fuera del entorno del OTS, en el proceso real, el riesgo limita ensayar este tipo de acciones.
- Se pueden establecer distintos puntos de partida para la sesión de entrenamiento: parada de emergencia, parada programada, arranque, baja carga o carga plena. La flexibilidad del OTS permite además guardar el avance de la simulación cuando se desee y utilizarlo como punto de partida en sesiones de entrenamiento posteriores.

- Todos los operadores de consola pueden practicar sobre eventos reales e hipótesis operativas. Los eventos son evaluados y ensayados en el simulador para luego desarrollar nuevos escenarios de práctica operativa en el OTS.
- Los instructores ven aliviada su tarea al no necesitar continuar con extensas y repetidas charlas para explicar el funcionamiento de un proceso y las acciones que puede demandar. Horas invertidas en múltiples sesiones, el agotamiento de instructor y el operador en formación; todo superado por la ejecución de un escenario operativo durante una breve sesión de entrenamiento en el simulador. El OTS ofrece una clara mejora en la eficiencia de la capacitación además de la evaluación inmediata de la eficacia de la actividad.
- El entrenamiento se realiza fuera de línea y en menor tiempo. Los tiempos de espera pueden reducirse al incrementar la velocidad de la simulación. También es posible ralentizar un proceso para poder visualizar y entender las acciones de, por ejemplo: acciones de interlocks, lazos de control complejos u otro proceso complejo.
- Permite realizar el análisis de distintos escenarios operativos. Es un entorno seguro para poner a prueba cambios de las variables operativas y evaluar los efectos en el proceso, determinar las limitaciones y poder evaluar la eficacia de los cambios realizados en los procedimientos.

"Si escucho olvido,
si veo recuerdo,
si hago comprendo"

CONFUCIO

Figura 11. Complejo Industrial Plaza Huincul, Neuquén, Argentina.

7 Referencias:

[1] OTS Especificación Funcional de Diseño. Abril 16, 2014. Invensys Operations Management. Buenos Aires, Argentina.

[2] Manual del Instructor OTS YPF Plaza Huincul – Metanol. Mayo, 2015. SimSci by Schneider Electric.

[3] DYNSIM® 5.2 Dynamic Simulation Suite User Guide, 2014. Invensys Systems, Inc. Rancho Parkway South Lake Forest, CA, U.S.A.

[4] M. McCall, M. M. Lombardo y R. A. Eichinger, 70:20:10 Model for Learning and Development (1996), Center for Creative Leadership.

[5] Entrevistas a usuarios del OTS – YPF. Planta Metanol – CIPH.

616513 Maximizing Returns from Enterprise Manufacturing Intelligence (EMI) and Multivariate Statistical Process Control (MSPC)

Abstract:

Enterprise Manufacturing Intelligence (EMI) has found great value in the chemical industry for aiding in timely decision making for improved plant reliability. It typically involves the use of control charts of multiple variables; that is, a univariate approach to data analysis. Another approach is to consider the data all together in a multivariate model, known as multivariate statistical process control (MSPC). Univariate and multivariate statistical process control are complementary techniques that can deliver greater value than either approach alone. Two use cases will be discussed including guidelines for maximizing returns from EMI and MSPC deployments, considerations when setting up the MSPC model and examples for how to interpret MSPC alerts.

619061 Data-Driven Identification of an Industrial Reverse Osmosis Membrane Process: Impact of the Feed Properties, Plant Variables and the Operating Window

Abstract:

The petroleum facilities require large amount of water for the cooling, boiler feed, processing, sanitary services, fire protection, and many other miscellaneous purposes. Both feed water and resultant water in the refineries must be well treated due to corrosive effect, quality requirements and environmental issues. Treating the water via reverse osmosis (RO) membranes is an important industrial process in terms of sustainability ([1]). This study focuses on the data driven modeling of an industrial Reverse Osmosis Membrane Unit (ROMU) which includes cartridge filters, high pressure pump, and RO membrane series. RO membranes are composed of two stages with 36 and 18 membranes respectively. In more detail, one membrane module contains 6 spiral wound polyamide membranes as shown in Fig. 1.

<![if !vml]>

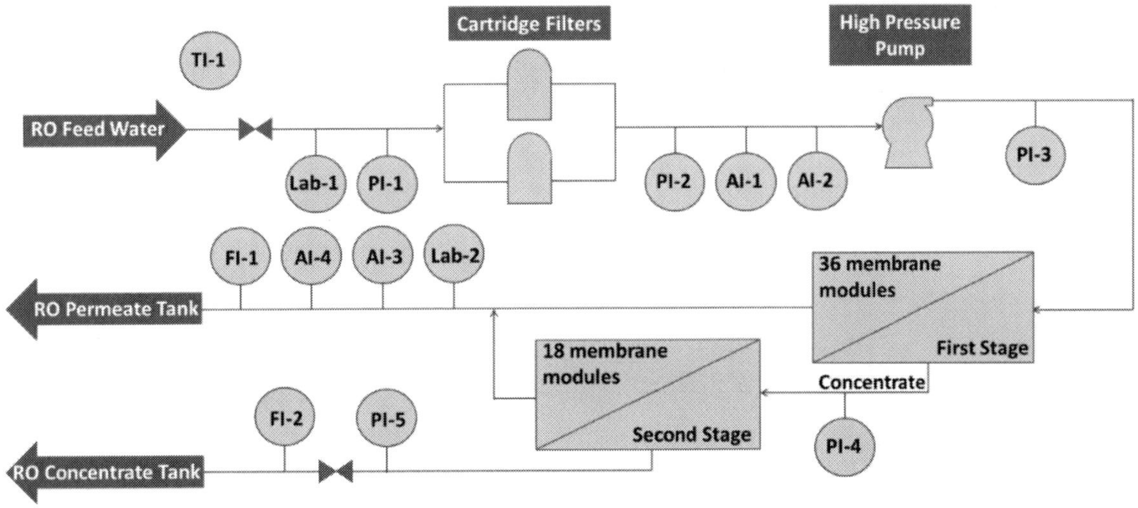

<![endif]>

Figure 1. Simplified ROMU diagram

Despite their wide usage, RO membranes suffer from permanent fouling once they are not maintained properly, resulting in higher cost and reduced product amount ([2]). In addition, the actual nature of the process is complex with high number of non-linearly coupled process variables. RO membrane fouling modeling and the determination of the major impacts on the process are the primary objectives of this study. Artificial Neural Network (ANN) is trained for the identification of the process to infer the impact of the plant and feed parameters on the outputs, primarily the product rate. The data for the model development contain the last 4 years, since the installation of the plant to date. The data include all sensors related to pressure, flow, temperature, conductivity, and pH. Next, the ANN is evaluated at probable operating conditions to calculate the operating window of the plant. The results show the impact of the process parameters on the product rate.

References

[1] Ismail, F., Khulbe, K. C., & Matsuura, T. (2018). *Reverse Osmosis*. Elsevier.

[2] Kang, G. D., & Cao, Y. M. (2012). Development of antifouling reverse osmosis membranes for water treatment: a review. *Water research*, *46*(3), 584-600.

619266 Elucidating Diafiltration Membrane Cascade Design Rules with Superstructure Optimization

Abstract:

The synthesis of many advanced materials in the past three decades have failed to produce a corresponding revolution in sustainable separation applications.[1] For example, new technologies for carbon capture can take up to 30 years to progress from a laboratory to commercial deployment.[2] To accelerate technology development, there is a critical need to move away from classical Edisonian iterative methods of materials discovery. Principled and data-driven frameworks will overcome the gaps in knowledge which inhibit the translation of new materials and devices from laboratory scale to sustainable separation technologies. In this talk, we describe our work using superstructure optimization at the systems scale and nonlinear parameter estimation at the materials scale in designing next generation membrane technologies and materials. We highlight how the current work lays the foundation for a holistic molecules-to-infrastructure computational design framework for the development of sustainable separation technologies.[3]

Relative to conventional technologies, membrane separations have demonstrated significant advantages in sustainability and energy efficiency.[4] For example, chemically selective separations enabled by polymer membranes have the potential to replace hazardous and toxic solvents used in conventional lithium-ion battery (LIB) recycling technology. However, there is no clear understanding of the interfacial and thermodynamic phenomena underlying the chemically-selective transport mechanisms needed to separate molecules with comparable sizes such as lithium (Li) and cobalt (Co) which hinders the development of membrane materials for this task.[5] Another challenge is the recovery of high purity products from a membrane separations unit. The formation of salt precipitate layers at the membrane interface when the feed side solution approaches the solubility limit prevents the extraction of high purity products, which also prevents the direct staging of these units.

In this talk, we demonstrate how a continuous, membrane based diafiltration cascade[6] enables the recovery of high purity Li and Co rich solutions for a LIB recycling case study at purities comparable with existing technologies. In diafiltration, the effects of salt precipitation encountered while staging membrane units are offset by adding a pure or dilute solvent, known as the dialysate to the feed side of the membrane filtration process. We show how the unique configuration of our system enables the utilization of the retentate side product, which is dilute in Li, to be recycled as dialysate to preceding stages of the process. Specifically, we highlight how superstructure optimization searches multiple, complex process configurations postulated by the modeler to identify promising applications for emerging materials. We treat flow and concentration of all streams as decision variables in our nonlinear optimization model. We then characterize the Pareto set using an epsilon-constrained method to understand the tradeoffs between purity and recovery objectives for the diafiltration cascade. Finally, we use the top-down capabilities of this framework to identify property targets for emerging membrane materials for LIB recycling.

Through our multi-objective superstructure optimization framework, we assemble a library of over 10,000 optimal designs spanning diverse system recovery specifications, area constraints, and membrane properties. From this library, we generalize common structures in process configuration into (near) optimal diafiltration cascade design rules. Many of these rules are non-trivial but ultimately justifiable by the governing transport and thermodynamic phenomena. In fact, some of these design rules show strong analogies between membrane diafiltration and distillation.

References

[1] Luo, J. & Crittenden, J. C. Nanomaterial Adsorbent Design: From Bench Scale Tests to Engineering Design. *Environmental Science & Technology* **53**, 10537 - 10538 (2019). 10.1021/acs.est.9b04371.

[2] Miller, D. C., Syamlal, M., Mebane, D. S., Storlie, C., Bhattacharyya, D., Sahinidis, N. V., Agarwal, D., Tong, C., Zitney, S. E., Sarkar, A., Sun, X., Sundaresan, S., Ryan, E., Engel, D. & Dale, C. Carbon Capture Simulation Initiative: A Case Study in Multiscale Modeling and New Challenges. *Annual Review of Chemical and Biomolecular Engineering* **5**, 301-323 (2014). 10.1146/annurev-chembioeng-060713-040321.

[3] Eugene, E. A., Phillip, W. A. & Dowling, A. W. Data science-enabled molecular-to-systems engineering for sustainable water treatment. Current Opinion in Chemical Engineering 26, 122-130 (2019). 10.1016/j.coche.2019.10.002.

[4] Phillip, W. A., & Elimelech, M. The Future of Seawater Desalination: Energy, Technology, and the Environment. *Science*, **333**, 6043, 712–717 (2011). 10.1126/science.1200488.

[5] Park HB, Kamcev J, Robeson LM, Elimelech M, Freeman BD. Maximizing the right stuff: the trade-off between membrane permeability and selectivity. *Science* **356**, 1-10 (2017). 10.1126/science.aab0530.

[6] Eugene, E., Phillip, W. & Dowling, A. Material Property Goals to Enable Continuous Diafiltration Membrane Cascades for Lithium-ion Battery Recycling. In *Computer Aided Chemical Engineering* **47**, 469-474 (2019). 10.1016/B978-0-12-818597-1.50075-8.

619247 Dynamic Modeling and Nonlinear Parameter Estimation of Nanofiltration Membranes

Abstract:

Compared to traditional chemical separation techniques, membrane processes offer compact configurations and energy-efficient means to achieve molecular scale separations. Membranes capable of achieving higher degrees of selectivity relies on both the precise control of a membrane's nanostructure and the identification of chemical functionalities which accentuate desired transport properties. [Hoffman & Phillip, 2020; Sadeghi et al., 2018] We argue process systems engineering offers untapped capabilities to balance membrane design decisions across (macro)molecular, device, and systems scales, and such a multiscale perspective is essential to fully realize the potential of nanoengineering to deliver transformative membrane separation technologies. [Eugene et al., 2019a] For example, we recently developed a superstructure optimization framework to design diafiltration membrane cascades. Our analysis shows that, through optimal systems-scale design, existing membrane materials have the potential to outperform current Li/Co separation methods for lithium-ion battery recycling; in other words, this particular application is mostly systems, not materials, limited. [Eugene et al., 2019b] In another study, we developed a process targeting framework to identify the necessary molecular and materials properties to realize absorptive membranes for environmental remediation. [Eugene et al., 2020] But extending these PSE paradigms to more directly inform nanoengineering requires more detailed and realistic mathematical models. [Geise et al, 2014; Yaroshchuk et al., 2018] Often the iterative creation and validation of said models is time-consuming and can be a rate-limiting step in scale-up, design, and analysis.

In this talk, we present an integrated experimental and data analysis technique to dynamically characterize nanofiltration membranes, resulting in differential-algebraic process models with quantified uncertainty suitable for process design. Experiments were run in a 10 mL dead-end filtration Amicon 8010 stirred cell connected with a dialysate tank. We conducted transport studies by loading both the cell and tank with predetermined concentrations of the desired salt and pressurizing the system with nitrogen gas. This apparatus design allows for continuous monitoring of permeate sample mass. Both retentate and sample concentrations are measured at regular intervals. We then postulate a family of differential-algebraic equation models for the system. Nonlinear regression is used to successfully calibrate hydraulic permeability, solute permeability coefficient, and reflection coefficient parameters in the model. These parameters correspond to the membrane transport properties and depend on the membrane chemistry, membrane nanostructure, and thermodynamics of the membrane-solution interface. As proof of concept, we consider diafiltration experiments for Mg^{2+} and K^{+} ions across an NF-90 nanofiltration membrane. We cast nonlinear parameter estimation as a multi-objective least-squares optimization problem and explore the trade-offs between different experimental observations, i.e., the mass of collected samples and sample concentrations. Through this analysis, we determined reflection coefficient is not identifiable in conventional filtration experiments and used the model to inform the diafiltration experiment design that covers higher concentrations. Finally, we discuss how the proposed dynamic experiments and data analysis technique require much less data than classical steady-state experiments to create and validate a model.

References

Eugene, E., Phillip, W., and Dowling, A., Data science-enabled molecular-to-systems engineering for sustinable water treatment, *Current Opinion in Chemical Engineering*, **26**, 122-130, (2019a).

Eugene, E., Phillip, W., and Dowling, A., Material Property Goals to Enable Continuous Diafiltration Membrane Cascades for Lithium-ion Battery Recycling, *Computer Aided Chemical Engineering,* **47**, 469-474, (2019b).

Eugene, E., Phillip, W., and Dowling, A., Material Property Targets for Emerging Nanomaterials to Enable Point-of-Use and Point-of-Entry Water Treatment Systems, Preprint, https://doi.org/10.26434/chemrxiv.12526190, (2020).

Geise, G. M., Paul, D. R., Freeman, B. D., Fundamental water and salt transport properties of polymeric materials, *Progress in Polymer Science*, **39**(1), 1-42, (2014)

Hoffman, J. R., and Phillip, W. A., 100th Anniversary of Macromolecular Science Viewpoint: Integrated Membrane Systems, *ACS Macro Letters*, 9, 1267-1279, (2020).

Sadeghi, I., Kaner, P., and Asatekin, A., Controlling and Expanding the Selectivity of Filtration Membranes, *Mater.*, **30**(21), 7328–7354 (2018).

Yaroshchuk, A., Bruening, M. L., and Zholkovskiy, E., Modelling nanofiltration of electrolyte solutions, *Advances in Colloid and Interface Science*, **268**, 39-63, (2019).

620535 Combining Data-Driven and First Principles Models: A Tutorial

Abstract:

Integrating physical, biological, chemical knowledge and "machine learning" is a critical aspect of developing industrially focused digital twins for monitoring, optimization, and design of (bio)chemical processes. However, identifying the correct model structure (e.g. the data-driven and the knowledge driven components to quantify the complex mechanisms) poses a severe challenge.

In this tutorial we will look at different hybrid modeling architectures, the variety of data-driven models that can be employed, as well as their benefits, and some insights into how to determine what is the best structure for a hybrid model.

622134 Dyos: A Software Package for Dynamic Optimization Using Direct Shooting

Abstract:

Over the past decades, dynamic optimization has become a fundamental instrument in process design and operation. We present DyOS [1], our in-house open-source software framework for solving optimization problems with large-scale differential algebraic equation systems (DAEs). DyOS features state-of-the-art open-source as well as commercial DAE integrators and nonlinear programming solvers. A special feature is that the parameterization of the control variables can be determined automatically using control grid adaptation [2, 3]. This allows for a highly performant problem-adapted solution of the dynamic optimization problems. In addition to various model interfaces for C++ formulations, DyOS provides an interface to functional mock-up units (FMU). FMU offer a standardized interface to higher object-oriented, declarative languages such as Modelica. Equipped with MATLAB, Python, and C++ bindings, optimizing with DyOS is straightforward. Moreover, a graphic user interface allows direct and intuitive usage. In this tutorial session, we give a brief overview of the features of DyOS and provide instructions on how to get started with the software. Here, we show how to set up and solve a simple nonlinear dynamic optimization problem. Finally, we demonstrate the strong capabilities of DyOS on small and large-scale real-world problems.

* Corresponding author: A. Mitsos

AVT Process Systems Engineering, RWTH Aachen University, 52056 Aachen, Germany

E-mail: amitsos@alum.mit.edu

Acknowledgements

The authors gratefully acknowledge the financial support of the Kopernikus project SynErgie by the Federal Ministry of Education and Research (BMBF) and of the KoPPonA 2.0 project by German Federal Ministry for Economic Affairs and Energy (BMWi) under grant number 03EN2004L and the project supervision for both projects by the project management organization Projektträger Jülich.

References

- Caspari, A.; Bremen, A. M.; Faust, J. M. M.; Jung, F.; Kappatou, C. D.; Sass, S.; Vaupel, Y.; Hannemann-Tamàs, R.; Mhamdi, A. & Mitsos, A., DyOS – A framework for optimization of large-scale differential algebraic equation systems, *Computer Aided Chemical Engineering*, 2019, 46, 619-624
- Schlegel, M.; Stockmann, K.; Binder, T. & Marquardt, W., Dynamic optimization using adaptive control vector parameterization, *Computers & Chemical Engineering*, **2005**, *29*, 1731-1751
- Assassa, F. & Marquardt, W., Optimality-based grid adaptation for input-affine optimal control problems, *Computers & Chemical Engineering*, **2016**, 92, 189-203

620535 Combining Data-Driven and First Principles Models: A Tutorial

Abstract:

Integrating physical, biological, chemical knowledge and "machine learning" is a critical aspect of developing industrially focused digital twins for monitoring, optimization, and design of (bio)chemical processes. However, identifying the correct model structure (e.g. the data-driven and the knowledge driven components to quantify the complex mechanisms) poses a severe challenge.

In this tutorial we will look at different hybrid modeling architectures, the variety of data-driven models that can be employed, as well as their benefits, and some insights into how to determine what is the best structure for a hybrid model.

619912 Establishing a Global Robotics Footprint & Its Continuous Evolution

Abstract:

Beyond the adoption of an emerging technology, the success and complexity of scale-up is often hindered by the value capture of the technologies. Dow will share how they established their Robotics globally by establishing a culture change in safety thru technology, using data-driven approach for trial & implementation of new platforms and reshaping our teams from implementation to development by integrating the technology into our work processes for maintenance and capital projects.

620019 Robotic Systems for Tank Floor Inspection

Abstract:

The storage tanks used to store crude oil, intermediates and fuels are subject to degradation because of the exposure to chemicals and the environment. Inspection processes are used to understand the condition of these tanks and plan for repairs or replacement. The floor of the tank is a major concern as it is in contact with soil, the product and it is not accessible when it is in service. Removing the tanks from operation, cleaning and performing inspection is a dangerous, costly and in many cases an unnecessary activity as the floor may have lasted longer. Knowing the condition of the floor with no disruption of operation is a very desirable solution.

For many years Chevron has been working with two different companies to develop these type of robotics. These systems have been tested and evaluated to comply with industry standards and our own requirements. They are now commercially available and we are in the process to integrate them in our workflows.

I will introduce the challenges and requirements for these systems, some of the business cases, the development work and the results obtained during these trials.

621792 Sponsored Technology Workshop - How to Leverage Physics-Based Simulation and AI/ML for Real Time Operations - Siemens Digital Industries Software

Abstract:

The use of a physics-based "Digital Twin" is getting more and more traction to develop process equipment virtually. Digital twins allow engineers to find the optimal design and operating parameters before the unit goes into production. However, these digital twins can't be deployed at the operational level because they can be complex or too slow to react at the speed of operation.

Join us for this workshop as we discuss a new set of solutions that lowers the barrier in executing these digital twins on the production floor. This will deliver substantial ROI for the chemical process industry.

In this workshop, you'll learn:

• The key requirements for physics-based simulation to be valuable

• Where the disconnect is between simulations done within R&D and what Operations need

• A methodology to overcome the gap between R&D and Operations

• Siemens physics-based digital twin capabilities dedicated to the chemical process industry

• How to leverage state-of-the-art Machine Learning techniques to develop real-time virtual sensors based on measured data and physics-based models

620955 Panel Discussion: Data Literacy in Chemical Industry

Abstract:

With Industry 4.0 and digitalization efforts transforming the chemical industry, data is available in high quantities, yet there is a skill gap in the workforce to deal with these mass amounts of data. Building data literacy skills enables the effective use of data for business outcomes. Gartner defines data literacy as "the ability to read, write and communicate data in context, including an understanding of data sources and constructs, analytical methods and techniques applied — and the ability to describe the use case, application, and resulting value." In this panel we will discuss how the chemical industry is finding a path to data literacy for its workforce, how to promote a culture of data literacy, how to overcome data literacy challenges, and the role of universities in enhancing data literacy skills in the workforce.

620258 Real-time analytics solution for augmented intelligence in ethylene

Abstract:

An Intelligent Real-Time Analytics Solution utilizes advanced analytics and AI/ML techniques to provide real-time insights. Customized to operations requirement, the solution with specific analytical features addresses essential needs including live gap analysis, improvement opportunities, instrument drifts, stability, performance benchmarks, predictive insights, and serves as a communication tool between engineers and operations. It also tracks performance and provide insights on a live basis for improving yield, efficiency and asset effectiveness.

A real-time solution, developed with a state-of-art technology and advanced analytics application for ethylene process industry results in several benefits.

- Benefits for Operators:
  - Provides more insightful information and prescriptive advice to make right decisions in a timely manner, avoid suboptimal operation and achieve significant profit improvement.
  - Provides a digitalized environment to address dynamic challenges and support safe and efficient operation
- Benefits for Engineers:
  - Data analytics provide a higher level of insights and understanding of process and equipment conditions
  - Helps engineers reduce data handling and model maintenance efforts and focus on plant improvement and troubleshooting

This paper discusses the various opportunities in the ethylene process and improvements possible from application of AI/ML solutions. The best methodology to address challenges in achieving the desired objectives and overall approach to ensure success.

SPRING21 +17TH GCPS
A Joint AIChE and CCPS Meeting

Got Dust?
Performing a Dust Hazard Analysis (DHA)

Alfonso Ibarreta, Timothy Myers, Michael Stern, Sean O'Hern, Achim Wechsung
Exponent, Inc.
1075 Worcester Street, Natick, MA 01760
aibarreta@exponent.com, tmyers@exponent.com, mstern@exponent.com, sohern@exponent.com, awechsung@exponent.com

Prepared for Presentation at
American Institute of Chemical Engineers
2021 Spring Meeting and 17th Global Congress on Process Safety
Virtual
April 18 - 22, 2021

Got Dust?
Performing a Dust Hazard Analysis (DHA)

Alfonso Ibarreta, Timothy Myers, Michael Stern, Sean O'Hern, Achim Wechsung
Exponent, Inc.
1075 Worcester Street, Natick, MA 01760
aibarreta@exponent.com

Keywords: DHA, Dust Hazard Analysis, Combustible Dust

Abstract

Combustible dust hazards can be present at industrial facilities that handle or produce finely divided solid particulates (aka dust). National Fire Protection Association (NFPA) standards (including NFPA 652 *Fundamentals on Combustible Dust*) require facilities to perform dust hazard analyses (DHAs) on new and existing processes handling combustible particulate solids. The facility owner or operator is responsible for determining whether the materials being handled are combustible or explosible, and if so, characterizing their properties as required to support the hazard analysis.

The DHA is intended to systematically evaluate the fire, deflagration, and explosion hazards of a facility and the mitigation methods present in the facility. The DHA must be performed or led by a qualified person. Depending on the resources of the company, it may be performed by in-house staff, or may be led by one or more outside subject matter experts.

This presentation will outline best practices for performing DHAs, including how to best identify combustible dust hazards in facilities. The presentation will include discussion of the following topics:

- Identifying relevant combustible dust standards
- Identifying testing data required for material evaluation
- Selecting a hazard analysis methodology
- Identifying hazard mitigation strategies

1 Introduction

Combustible dust hazards can be present at industrial facilities that handle or produce finely divided solid particulates (e.g., powders or dust). Some industrial facilities intentionally handle powders as a raw material, intermediate, or final product. Other facilities generate dust as a byproduct or waste stream during operations such as cutting, grinding, or abrasive blasting.

National Fire Protection Association (NFPA) standards (including NFPA 652 *Fundamentals on Combustible Dust* [1]) require facilities to perform Dust Hazard Analyses (DHAs) on new and existing processes handling combustible particulate solids. Aside from the normal fire hazard posed by all combustibles, the unique hazards posed by a combustible dust are the following:

- Flash Fire: "A fire that spreads by means of a flame front rapidly through a diffuse fuel, such as dust, gas, or the vapors of an ignitable liquid, without the production of damaging pressure." [1]
- Explosion: "The bursting or rupture of an enclosure or container due to the development of internal pressure from a deflagration." [1]
    - Where 'deflagration' is defined as "Propagation of a combustion zone at a velocity that is less than the speed of sound in the unreacted medium." [1]

A combustible dust explosion hazard exists when all 5 elements of the explosion pentagon, shown in Figure 1, can be present at one time. If confinement is not present, a flash fire hazard could still be present. The goal of the DHA is to identify potential combustible dust hazards. The DHA team may also propose recommendations for how to prevent or mitigate hazards identified to have insufficient safeguards in place.

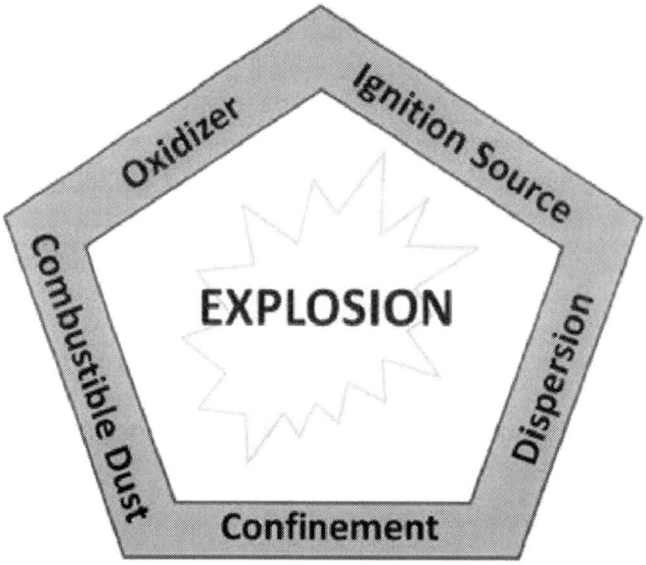

Figure 1. Combustible dust explosion pentagon showing the five elements necessary for an explosion to occur.

The NFPA standards that govern DHAs permit the owner/operator to follow multiple methodologies and strategies. The purpose of this paper is to summarize the DHA requirements and provide general guidance to owners and operators in how to prepare for and perform a DHA.

2 Combustible Dust Standards

The NFPA has developed a series of standards for the prevention and mitigation of dust explosions across a wide range of industries.[1] The NFPA has no enforcement authority, but these standards become regulations when adopted by an authority having jurisdiction (AHJ) such as when adopted by building codes, local, and state regulations. They are also typically considered good engineering practice, even when not required by regulations. NFPA combustible dust standards have been cited by OSHA when issuing citations per OSHA's General Duty Clause.

NFPA 652 *Fundamentals of Combustible Dust* provides a set of general requirements for the mitigation of fire and explosion hazards associated with combustible dusts and particulate solids

[1] The NFPA technical committees are currently working to consolidate these standards into a single document, NFPA 660.

across all industries and processes. The standard also directs the user to other commodity-specific NFPA combustible dust standards for additional requirements. These include:

- NFPA 61, *Standard for the Prevention of Fires and Dust Explosions in Agricultural and Food Processing Facilities*
- NFPA 120, *Standard for Fire Prevention and Control in Coal Mines*
- NFPA 484, *Standard for Combustible Metals*
- NFPA 654, *Standard for the Prevention of Fire and Dust Explosions from the Manufacturing, Processing, and Handling of Combustible Particulate Solids*
- NFPA 655, *Standard for Prevention of Sulfur Fires and Explosions*
- NFPA 664, *Standard for the Prevention of Fires and Explosions in Wood Processing and Woodworking Facilities*
- NFPA 850, *Recommended Practice for Fire Protection for Electric Generating Plants and High Voltage Direct Current Converter Stations*

These standards provide requirements to prevent the occurrence of ignitable dust clouds in the presence of ignition sources. Additionally, the standards contain requirements for mitigation methods that minimize the effect of dust explosions or fires if they do occur. A DHA will, in part, check for compliance with the applicable NFPA standards.

3 Determining Whether Your Dust is Combustible

Whether or not a Dust Hazard Analysis (DHA) needs to be performed in the first place depends on whether the dust produced or handled at the facility is combustible and/or explosible. According to NFPA 652, a particulate solid is defined as 'combustible' if it can propagate a flame or smoldering front when a 25cm-long train of accumulated material is subjected to a flame, and 'explosible' if it is "capable of propagating a deflagration when dispersed in air or the process-specific oxidizing media."[1] The facility owner or operator is responsible for determining whether the materials being handled are combustible or explosible, and if so, for characterizing their properties as required to support the hazard analysis.

While NFPA 652 permits the use of historical facility data or published data to determine whether a dust is combustible or explosible, it is important to note that variations in particle size and morphology, moisture content, composition, or other properties could affect the combustibility and explosibility of a material. Testing of a representative sample using the Combustibility Screening Test (Trough Test) and the Explosibility Screening Test (Go/No Test) is typically the preferred methodology [1]. The same material in different locations of the facility may exhibit different combustibility behavior. For example, plastic resin pellets used as a feedstock may not be combustible but small particulates of that resin accumulated on elevated surfaces or on vacuum pump filters may be combustible. Therefore, appropriate sample selection and/or sieving the sample to test the fine fraction of a sample are important for accurately assessing whether a combustible dust is present.

Once a dust is classified as combustible and/or explosible, additional tests may be required to analyze the potential hazard and determine the level of protection required. Some of the parameters that can be quantified include [2]:

- Deflagration Index (K_{st})
- Maximum Pressure (P_{max})
- Minimum Explosible Concentration (MEC)
- Limiting Oxygen Concentration (LOC)
- Minimum Hot Surface Ignition Temperature (MIT)
- Minimum Ignition Energy (MIE)
- Minimum Autoignition Temperature (MAIT)

Determination of the minimum ignition energy, minimum autoignition temperature, and hot surface ignition temperature are useful for determining the likelihood of ignition. Based on these results, safe operating conditions, such as maximum equipment temperatures, can be established. A material with a low minimum ignition energy may be susceptible to sparks and static discharges that necessitate certain precautions in handling.

The P_{max} and K_{st} values quantify the severity of a potential explosion and can be used to design explosion protection vents or chemical suppression systems. In some equipment or processes where practical, inert gas may be used to lower the oxygen content in the enclosure. Measurement of LOC value allows designers to determine the maximum permissible levels of oxygen in order to sufficiently inhibit deflagrations. Additional test methods are available for determining other properties of the dust.

It is typically not necessary to perform all tests on all dust samples, but rather specific properties can be measured as needed to perform the DHA or design protection systems.

4 Performing a Dust Hazard Analysis (DHA)

NFPA 652 requires that a DHA be performed at any facility handling combustible dust, and that the DHA be reviewed and updated every 5 years [1]. In addition, changes in equipment or operating conditions or materials should trigger a management of change review, which would address any hazards that are introduced by the changes [1]. Therefore, a hazard analysis is a necessary step in the implementation of any new or updated process that handles combustible dust.

4.1 DHA Methodology

According to NFPA 652, "the DHA shall evaluate the fire, deflagration, and explosion hazards and provide recommendations to manage the hazards" [1]. However, NFPA 652 does not prescribe the evaluation methodology and allows the owner/operator to identify and utilize which methodology or methodologies to best meet their needs. A wide range of hazard analysis strategies are available, including:

- Hazard and Operability Study (HAZOP)
- Checklist

- What-If Analysis
- Failure Modes and Effects Analysis (FMEA)
- Event Tree Analysis
- Fault Tree Analysis
- Cause-Consequence Analysis
- Bow-Tie Analysis

These and other methodologies are described in detail in the Center for Chemical Process Safety books, *Guidelines for Combustible Dust Hazard Analysis* 1st edition [3] and *Guidelines for Hazard Evaluation Procedures*, 3rd edition, published by the American Institute of Chemical Engineers [4]. It should be noted that it is not the intent of the NFPA 652 standard to require owners and operators to follow Process Safety Management (PSM) as defined in OSHA regulation 29 CFR 1910.119 [5] in performing the DHA [1]. However, some of the elements of PSM are included in NFPA 652, and some dust handling processes may fall under PSM.

The DHA must be performed or led by a qualified person [1]. The qualified person should be familiar with conducting a DHA, as well as with combustible dust hazards. The DHA is normally performed by a team, which should include persons that are familiar with the process operations (e.g., engineers, operators, maintenance, safety).

4.2 Hazard Mitigation and Prevention Strategies

As described above, the DHA evaluates the presence of hazards and the current safeguards designed to prevent or mitigate the hazard. Safeguards are required to ensure an appropriate level of risk to life safety but may also be warranted if the risk of operability disruptions or environmental releases exceeds the desired level. For example, an outdoor and remotely located dry material dust collector for non-metal combustible dust may not require explosion protection based on life safety considerations. Adding explosion protection, however, might significantly reduce downtime in the event of an internal deflagration within the collector.

Explosion mitigation and prevention systems function by eliminating one of the required conditions for a combustible dust explosion shown in Figure 1. The presence of combustible dust can sometimes be eliminated or reduced through substitution of the feedstock with a less combustible material or dilution of the combustible material with an inert material. The presence of an oxidizer can sometimes be eliminated or reduced through addition of an inert gas to partially displace the oxygen-containing air within equipment. Confinement, which is required for a deflagration to result in an explosion, can be reduced by adding explosion vents. Dispersion of the combustible dust can be reduced through the addition of dust collection systems and housekeeping procedures.

NFPA 652 provides multiple means to assess the compliance of the system being reviewed. Compliance can be assessed via a prescriptive DHA, a performance-based DHA [1], or a risk-based DHA. In a prescriptive DHA (such as a "checklist" analysis), the owner/operator ensures that all of the equipment, processes, and management systems meet or exceed each requirement listed in the appropriate chapters of NFPA 652 and any commodity-specific NFPA standards. A performance-based DHA evaluates meeting the performance requirements outlined in Chapter 6 of NFPA 652 and may not meet each outlined prescriptive requirement from other chapters. Furthermore, NFPA 652 permits the prescriptive requirements to be modified by a documented risk analysis, which allows for further flexibility in the meeting of the DHA objectives. Typically, a hybrid approach is used where most of the prescriptive requirements of the standard are followed, but where unique issues arise, a performance-based or risk-based analysis is performed.

Following the prescriptive requirements is one method to assess whether the risk of a combustible dust incident is at an acceptable level based on the properties of the dust and the safeguards present. In some cases, however, some or all the prescriptive requirements are unfeasible or undesirable for a given system. Alternatives to the prescriptive requirements may be permitted by the NFPA standards if they are an Equivalent Design, a Performance-Based Design, or based on a Risk Evaluation [6].

Performance-Based Designs must be evaluated by a qualified person based on design objectives including life safety objectives set forth in the NFPA standards and other criteria for operability acceptable to the owner/operator. The design objectives are then used to define performance criteria that ultimately determine the equipment and procedural thresholds and conditions that must be met in order to meet the design objectives. The development of these conditions and thresholds and their evaluation must be documented.

One example of an alternative design would be the use of administrative controls to eliminate or significantly reduce the probability of an employee being near a hazard at the time of a combustible dust incident. This alternative could meet life safety objectives if the probability of an employee being near a hazard in the event of a fire or explosion were sufficiently small and if the possible disruptions caused by a loss of equipment were acceptable to the owner/operator.

4.3 DHA Recommendations

The results of the DHA review may involve recommendations to prevent or mitigate identified combustible dust hazards with insufficient safeguards currently in place. Recommendations should be actionable items that consider the feasibility and effectiveness of the proposed mitigation. Examples of recommendations often found in DHAs include:

- The facility needs to develop appropriate housekeeping procedures (including scheduling, regular inspections and unscheduled housekeeping procedures) to maintain the dust accumulations below 1/32 inch on horizontal surfaces, including elevated surfaces.
- The facility needs to develop a bonding and grounding auditing program to ensure that equipment remains properly bonded and grounded.
- The dust collector needs to be fitted with a properly-sized explosion vent that directs a potential deflagration to a safe outside location, in accordance with NFPA 68 (2018).

The owner / operator must document the results of the DHA and properly address the recommendations. Typically, there may be multiple methods to mitigate the risk of hazards identified by a DHA. Completing the DHA can be an iterative process as various solutions are evaluated until final responses to the DHA recommendations are prepared. Documentation of the response to the recommendations must include any necessary action items requiring change to the process materials, physical process, operating conditions, or facilities.

The owner / operator is responsible for maintaining appropriate management systems. This includes operating procedures and protocols, housekeeping procedures, hot work permits, inspection and maintenance programs, training and hazard awareness programs for personnel and contractors, emergency planning and response plans, and incident investigation systems. For each of these, a management of change process helps owners in considering safety implications before changes are made and a document retention program ensures that owners/operators keep proper records.

Lastly, the owner / operator is also responsible for reviewing and updating the DHA at least every five years or when material changes to the conditions occur.

5 Conclusions

NFPA 652 and other commodity-specific combustible dust standards require that a DHA be performed at facilities handling combustible dust, and that the DHA be reviewed and updated every 5 years. The first step in the process involves testing the particulate materials handled or produced at the facility, to determine if they are combustible and/or explosible.

The applicable NFPA standards allow the use of a variety of methods and strategies to ensure that combustible dust hazards are properly mitigated. Ultimately, it is the responsibility of the facility owner / operator to ensure the DHA is properly performed and the recommendations stemming from the DHA are addressed.

6 References.

[1] NFPA 652 *Fundamentals of Combustible Dust.* 2019 Edition.

[2] Michael C. Stern, Alfonso Ibarreta, Timothy J. Myers. *Assessment and Mitigation of Combustible Dust Hazards in the Plastics Industry.* 30th International Conference of the Polymer Processing Society (PPS-30), Cleveland, OH, June 8-12, (2014)

[3] Center for Chemical Process Safety (CCPS), "Guidelines for Combustible Dust Hazard Analysis." 1st Edition, Wiley & Sons, Inc., Hoboken, NJ (2017),

[4] Center for Chemical Process Safety (CCPS), "Guidelines for Hazard Evaluation Procedures", 3rd Edition, Wiley & Sons, Inc., Hoboken, NJ (2008),

[5] Code of Federal Regulations – Occupational Safety and Health Administration (OSHA): 29 CFR 1910.119 *Process safety management of highly hazardous chemicals.*

[6] Timothy Myers, Alfonso Ibarreta, Kevin Marr. "Prescriptive versus Performance-Based Mitigation of Combustible Dust Hazards" Proceedings of the American Institute of Chemical Engineers 2014 Spring Meeting, 10th Global Congress on Process Safety New Orleans, LA, March 30 – April 2 (2014).

620782 Practical Use Cases for Artificial Intelligence and Machine Learning (Poster)

Abstract:

As the digital transformation of industry progresses, an increasing amount of data is becoming available. The data only has value if it is analyzed and converted into useful information from which actionable decisions can be made. In parallel, the adoption of artificial intelligence (AI) technology has been increasing significantly. To date, the adoption of AI on the shop floor has been limited mostly to local feasibility studies that lack the ability to scale creating a hurdle for commercial adoption.

The question is, how can AI techniques such as machine learning capitalize on the newly available data sources to assist companies with the application of "industrial grade AI" on the shop floor in a scalable way that will address current maintenance and operations issues.

This presentation will include multiple use case examples of the application of AI in the factory to enable anomaly detection and data analytics through ML to predict failures, which decreases unexpected plant shutdowns, and reduce plant operation costs by enabling predictive maintenance and minimizing wear. Additionally, it will cover the use of AI to tie business targets to process control.

620615 Enabling Digital Transformation: It's about More Than the Analytics (Poster)

Abstract:

Over the past 5 years, the manufacturing industry has come to understand the potential impact of and the value of data analytics in the process environment. What followed was a deluge of vendors, each shilling their particular flavor of analytics – machine learning, multivariate, artificial intelligence, etc. In spite of the multitude of analytics options, most manufacturers continue to struggle with fully leveraging analytics output to deliver actionable insights in any scalable and sustainable way. As the speed of and emphasis on digital transformation increases, those struggles have been exacerbated.

Despite the growth of investments in individual analytics technologies, manufacturers are only experiencing marginal gains in their ability to analyze, understand and implement insights generated from the data. The manufacturers setting the pace in the industry today understand the need to move beyond the simple application of data analytics. They are instead starting to look at enabling their organizations with analytics knowledge.

Through analytics knowledge enablement, those insights can be leveraged anywhere in the company to anticipate and solve potential process upsets faster and with greater predictability. These leading manufacturers understand that generating the initial signal with analytics is only the first step. The larger value comes from the ability to then harvest and harness the knowledge generated across the organization.

This poster will:

Highlight the critical differences between simple analytics usage and manufacturing analytics knowledge enablement;
Demonstrate how manufacturing analytics knowledge enablement delivers competitive advantage;
And the requirements to deliver scalable and sustainable manufacturing analytics knowledge enablement.

Making Circular World Possible Through Chemistry

Circular Economy drives for consistent shift from linear economy system where we "take-make-use-dispose" to a system that utilizes reviewable energy, looks at how we can use things smarter and longer and at how do we design out the waste. As engineers and scientists, we need to leverage innovation and chemistry to design new technologies and products that enable more sustainable future and economy that works for our society in the long-term. Chemical Industries are in the middle of Circular Economy with manufacturing and processing of fossil fueled chemicals adding catastrophic emissions to our environment. To move towards more sustainable future, one of the largest areas of focus for chemical industries is recycling of legacy toxic chemical though chemical and mechanical recycling. Another significant focus is on enabling and designing technologies and products that contribute to lower carbon dioxide (CO_2) footprint, including utilization of renewable raw materials. However, there needs to be economical infrastructure that allows corporations to remain competitive while becoming sustainable. These efforts require planning, time and creativity.

At Rogers Corporation we are developing transformative ideas that help build cleaner, safer and more connected world. We demonstrate commitment towards reduction of global warming emissions in the products we sell that drive sustainable future and reduce dependence on fossil fuels. For instance, Rogers has served the renewable energy market for over 30 years with a focus on wind and solar, as well as power for general industrial applications and major appliances. Renewable energy is critical to reduce the greenhouse gas emissions and mitigate climate change. As another example, a number of our technologies are utilized as a component in the design of electric vehicles. According to the Environmental Protection Agency (EPA) estimates a typical passenger vehicle emits approximately 4.6 metric tons of CO_2 per year. Electrifying the vehicles and coming up with lightweight designs reduces emissions amount significantly. Our next generation products are focused on utilizing renewable resources and on recyclability. We are at the same time mindful of the environmental impact of our operational footprint, particularly as it relates to climate emissions and water, which we actively monitor and seek to reduce through our continuous improvement processes and recycling initiatives.

Circular economy is about transforming the entire value creation system to create a sustainable and low-carbon economy and society. In addition to converting to renewable energy, the next frontier for the chemical industries is engineering out "legacy chemicals" with toxic substances, or substances that cannot be recycled as well as designing products that last longer or can be repaired. Chemical Industries, such as Rogers Corporation can make a huge impact, but it will take all of us working together to give our future a fighting chance: governments, corporations and individuals.

Dr. Polina R. Ware

Global R&D Director

Rogers Corporation

01/25/2020

620469 Using Digital Process Twin Technology to Drive Operational Excellence in Utility Systems (Poster)

Abstract:

Digital replicas of operating assets that combine plant data with high-fidelity process models are bringing a new level of decision support to operations of steam, condensate, fuel gas and power systems. They provide many benefits, from monitoring and 'soft-sensing' to advice on optimal set-points, diagnostics and prognostics such as maintenance interventions. At the heart of such shadowing systems is an always-current digital simulation model of the assets which updates itself periodically based on real-plant performance.

In the context of utility systems such as steam, condensate and power, digital twin technology is now being implemented. The associated benefits include: equipment monitoring to determine the real state of equipment; real-time 'soft-sensing' to provide up-to-date performance information which is either difficult or impossible to measure; forecasting to determine future performance such as maintenance intervals based on current equipment state and anticipated operation; operational optimisation to give advice to operators regarding set-points, diagnostics, prognostics and performance benchmarks; and finally 'what-if' analysis to anticipate how to operate for future/alternative operating scenarios.

Successful digital operating systems use high-fidelity predictive mathematical models of the assets to exploit redundancy between model prediction and plant data and maintain themselves in an always-current state. This means that 'drift' in key parameters such as gas turbine efficiency, steam pressure levels, equipment fouling and so on is taken into account in all monitoring, forecasting and optimisation calculations, to provide reliable information for decision support.

This poster describes how the combination of equation-based general-purpose modelling technologies and next-generation digital application frameworks provide an environment for easy construction and application of fast, robust online solutions based on high-fidelity utility system models. A real industrial example is presented where the digital twin of a steam system is used to improve the system's performance. The twin is linked to plant data systems, updating itself through machine-learning capabilities, validating actual performance and, where appropriate, identifying departures from normal operation. Beneficial operating changes are highlighted. Visualisation of the twin is seen to be critically important to give operators greater insight and confidence to operate the process safely at the optimum point, thereby making a major contribution towards Operational Excellence.

620301 How To Ensure Safety-Critical Communication In Pandemic Times

Abstract:

The industrial internet of things (IIoT) has brought increasing emphasis on interconnecting sensors and devices in manufacturing facilities. Now applications utilizing cloud computing and AI are helping organizations leverage their IIoT infrastructure. This has help to move IIoT beyond the confines of IT to the shop floor, all to optimize efficiencies and productivity.

Even though this transformation is happening rapidly, it is vital not to overlook the *fact that manufacturing still needs people* (my, hasn't COVID proven this?). While IIoT enables faster and better reactions to operating conditions, even better trained staff are needed to understand and control these systems and to adapt them to changes more rapidly. Thus a "people-focused" IIoT enables industrial enterprises to harness true workforce know-how to maximize plant operations by increasing production efficiency which, most importantly, ensures employee safety.

Plant Process Management (PPM) systems help manage, monitor and optimize plants in addition to work activities, operations and production assets. They give a real-time view of work progress and track status of any incidents or outliers to routine operations. Current employees working in shift manufacturing need to be made aware of any issues that may have occurred in the previous shift, which isn't always easy. Poor communication can directly impact safety, quality and productivity.

PPM delivers on the requirements for all plant operations, regardless of the use of automation, IoT instrumentation, machine learning (ML) or predictive maintenance. Knowledge can be gained from any human contact point in a process. Consistent capture of team-to-team communication is critical, especially with changing work patterns due to working remotely or social distancing. For example, COVID-19 has put a strain on conducting safe operations but fortunately PPM is one solution that can safeguard manufacturers and further optimize their operations.

For most manufacturers, IT and OT represent sets of separate and distinct information silos, hindering the progress of IoT. PPM solutions help bridge this gap by allowing teams to seamlessly interact with OT data and IT applications, for example.

In a survey conducted by 451 Research, part of S&P Global Market Intelligence, more than 50% of OT respondents surveyed confided that they seldom work together with their IT team. Of those surveyed, only about 20% of IT leaders collaborate with OT. IT teams believe the key problem lies in the inability to share captured data where OT teams feel reconciling shift data with machine logs poses the biggest challenge. However, both IT/OT participants surveyed agreed the following points are a top priority:

- Capturing and sharing best practices for employee safety
- Implementing real-time incident tracking to conduct corrective action
- More cohesive collaboration and communication among IT/OT teams
- Automated interactive production loss accounting

According to research findings, a focus on worker safety is an important driver for changing the status quo. Without repeatable procedures in place where human communication occurs, a disconnect can occur where bad incidents could happen. As equipment and resources are run and maintained by a human workforce, the correct tooling and equipment, including those enabling clear communication and collaboration around details are essential.

It is possible to digitally transform operations with PPM, without a full IIoT roll-out. Consider quick wins such as digitizing shift handovers, safety and compliance tracking, and corrective maintenance planning. Or what about replacing whiteboards, ad hoc conversations and checklist sheets with an electronic system that anyone can access, anywhere? Once digitized with PPM, these common elements of everyday shift work can easily take advantage of future IIoT capabilities and become powerful assets in themselves to the production operation.

However, if the workforce is not taken into account in your digital transformation, potential benefits can be lost. In addition, as older workers retire and are not replaced as quickly as they leave, vital knowledge from their years of experience can disappear without a digital PPM-based knowledge base. Furthermore, the COVID-19 pandemic, as well as the potential for future crises (whether global or local), highlights the necessity for digital tools that help teams communicate, coordinate, and co-operate – wherever they happen to be.

Consider PPM as the next step in your path to digital transformation.

619300 Use of IEC 62443 Standards to Enhance Industrial / OT Cybersecurity in Chemical Industry

Abstract:

Chemical Industry Sector is one of the most important Critical Infrastructure Sectors as per the Executive Order promulgated by President Obama in February 2012. Like other sectors, this sector is highly vulnerable to Cyber and Cyber-Physical attacks.

With the recent spate of Cyberattacks on Chemical Industry, it is necessary to consider "Defense-in-Depth" approach to foster Cyberdefense for the Industry, especially augmenting Cybersecurity of Industrial Control System (ICS) in a Chemical Plant.

IEC 62443, which has replaced old and well-known standard ISA-99 is one of the ways to bolster OT or Industrial Cybersecurity, especially that of ICS which are extensively used in Chemical Industry to track, monitor and control important process parameters such as temperature, flow, pressure, PH etc.

This case-study based presentation gives comprehensive and good understanding and excellent insight into the following:

1) Brief overview of Cybersecurity Risk faced by the Chemical Sector as a part of Critical Infrastructure

2) Introduction to Operational Technology and Cybersecurity Challenges & Issues faced by OT / IIoT / ICS used in the Chemical Industry.

3) Brief overview of some of the well-know attacks including Ransomware attacks targeted to this Sector in last few years

4) IT and OT Convergence and Alignment to strengthen the Cybersecurity

5) Overview of IEC 62443 Series of Standards and their applicability across the entire value-chain and lifecycle of ICS and Process Control System in the Chemical Sector

6) Deep-dive into Purdue Model, Network Architecture and concept of Zones and Conduits

7) Implementation Roadmap to implement various IEC 62443 series of standards as applicable to different stakeholders

8) Certification Roadmap to IEC 62443

9) Case Study discussion

10) Recommendations for improving & strengthening OT / ICS Cybersecurity along with Safety in a typical Chemical Plant

11) Summary and Lessons learnt followed by Q & A

With the above detailed agenda, this presentation will be of tremendous value and very practical and useful resource for those practitioners and different stakeholders who are preparing and planning to address and mitigate Cybersecurity Risk and Challenges that Chemical Industry is currently facing especially to their Operational Technology, Industrial Control Systems (ICS), SCADA systems and so on.

623026 Overview of a Continuous Process Test Bed Being Built and the Mobile Worker Use Cases That Use Augmented Reality Backed By Digital Twin Technology (Poster)

Abstract:

In this abstract the audience will learn about augmented reality and digital twin use cases around a continuous process test bed being built by a representative group of industry players and stake-holders. This industrial skid will be available for industry professionals and students to visit at the MxD facility on Goose Island, Chicago in 2021. My proposal for this skid was selected for execution under funding from US Government backed `Manufacturing USA` initiative by the Manufacturing times Digitalization Innovation Institute and is currently at an advanced stage.
Background:

Manufacturing USA was created in 2014 to secure U.S. global leadership in advanced manufacturing by connecting people, ideas, and technology. Manufacturing USA institutes convene business competitors, academic institutions, and other stakeholders to test applications of new technology, create new products, reduce cost and risk, and enable the manufacturing workforce with the skills of the future (www.manufacturingusa.com).
MxD encourages factories across the United States to deploy digital manufacturing and design technologies, so those factories can become more efficient and cost-competitive (www.mxdusa.org).
Use Cases included in this abstract benefit from extended reality and digital twinning:
The goal of the process industry test bed skid (physical) and its digital twin (virtual) is to enable field operators, engineers, factory managers, and other key manufacturing roles to gain data-driven insights for increased operational awareness in order to better solve manufacturing problems and optimize processes. A factory manager should be able to communicate with and assign tasks to the mobile worker and view all data necessary to make informed decisions about the assignment of tasks. The solution will be demonstrated using the process test bed and should be agnostic to sensor type, analytics solution, and control type. Visualization of this data should include, but is not limited to, dashboards for factory managers and augmented reality technology for field operation using mobile devices (e.g. tablet). This project should answer the key question: How would you visualize sensor data with the best User Interface/User Experience design to get a seamless experience in continuous manufacturing?

619076 Practical Use Cases for Artificial Intelligence and Machine Learning

Abstract:

As the digital transformation of industry progresses, an increasing amount of data is becoming available. The data only has value if it is analyzed and converted into useful information from which actionable decisions can be made. In parallel, the adoption of artificial intelligence (AI) technology has been increasing significantly. To date, the adoption of AI on the shop floor has been limited mostly to local feasibility studies that lack the ability to scale creating a hurdle for commercial adoption.

The question is, how can AI techniques such as machine learning capitalize on the newly available data sources to assist companies with the application of "industrial grade AI" on the shop floor in a scalable way that will address current maintenance and operations issues.

This presentation will include multiple use case examples of the application of AI in the factory to enable anomaly detection and data analytics through ML to predict failures, which decreases unexpected plant shutdowns, and reduce plant operation costs by enabling predictive maintenance and minimizing wear.

618574 Top Five Criteria to Selecting a Successful AI Use Case

Abstract:

The potential of AI and machine learning is an exciting development for the chemicals industry. However, machine learning isn't the best solution for every challenge. In fact, estimated that by 2022, 85% of machine learning projects will fail – showing the need to have a clear understanding of Machine Learning and its suitability for your use case. In this keynote session, learn about the five important criteria for selecting a use case to ensure you start your AI strategy on the right foot.

By going through this process, you will be able to understand if machine learning is the best way to approach your pressing issues. Often, it will be. But if you throw ML at an arbitrarily chosen problem, there's no guarantee that it will be worth the investment.

619944 Democratizing Data Analytics for the Process Manufacturing Expert

Abstract:

The chemical industry has been confronted with innumerous challenges. Some of them have been boiling over the past few decades, such as increasing fluctuation in commodity prices, shrinking margins and ageing infrastructure. Covid-19 has brought to bear scenarios that were not as imminent, namely deglobalization and feedstock shortages. Amidst what some would call an unsettling situation, data analytics creates opportunities to offset these challenges. Leading chemical companies have pursued vertical data integration of their industrial (OT) and corporate (IT) systems. SAS enables those companies in their task to translate data into meaningful insights focusing on costs reduction, productivity improvement and market growth. In this presentation, SAS will show examples of success stories which touch upon different value drivers: OEE, energy efficiency, yield/throughput, and serviceability. In addition to that, SAS will also present lessons learned during its 44-year quest to democratize data analytics and make analytics accessible from the process engineer to the data scientist.

619989 Machine-Learned Decision Trees for Predicting Gold Nanorod Dimensions from Spectra Alone

Abstract:

Measuring the size of a single plasmonic nanoparticle is a critical step in correlating structure to optical properties but is often difficult to perform especially under irreversible conditions or non-immobilized nanostructures. Here, we demonstrate that gold nanorod dimensions can be accurately predicted over a wide range of aspect ratios using a simple machine learning model, decision trees. The model is trained using ~400 nanorod geometries and their corresponding scattering spectra obtained from finite-difference time-domain(FDTD) simulations. When validated on simulated FDTD spectra, 90 % of all predicted AuNR dimensions are within 10% relative error of the ground truth. Our results are tested using experimental spectra correlated with scanning electron microscopy images to obtain the sizes, estimating small and large nanorods dimensions with 8.5% relative error. Analysis of the decision tree structure reveals that a simple correlation with spectral peak position and linewidth of the localized surface plasmon resonance is adequate to predict nanorod dimensions, outperforming more complicated models. Our findings illustrate the advantages of using simple machine learning models to analyze single particle structural features from their optical spectra.

620017 Anomaly Detection in Process Data Using Generative Adversarial Networks (GAN)

Abstract:

The development of *DetlaV Advanced Continuous Historian* has paved the way for highly scalable storage of live plant data in the time series format. Conducting anomaly detection in such abundant process data can provide insights into irregular plant behavior and advise process engineers to take proactive measures. The recent proliferation of deep learning methods which excel in learning complex spatial/temporal correlations shows great promises in anomaly detection. In current work, we demonstrate the adoption of an unsupervised times series anomaly detection framework based on generative adversarial networks (GAN) [1, 2]. The efficacy and efficiency of the framework is illustrated using both open source data and industrial data. Such results demonstrate the successful application of state-of-art artificial intelligence to cope with real problems in chemical manufacturing within the Digital Transformation era.

[1] Geiger, A., Liu, D., Alnegheimish, S., Cuesta-Infante, A., & Veeramachaneni, K. (2020). TadGAN: Time Series Anomaly Detection Using Generative Adversarial Networks. *arXiv preprint arXiv:2009.07769*.

[2] Li, D., Chen, D., Jin, B., Shi, L., Goh, J., & Ng, S. K. (2019, September). MAD-GAN: Multivariate anomaly detection for time series data with generative adversarial networks. In *International Conference on Artificial Neural Networks* (pp. 703-716). Springer, Cham.

618013 How Well CAN WE Predict Causes behind the Pipeline Incidents?

Abstract:

Causation analysis of pipeline failure using past incident data has been in practice for direct causes, such as corrosion, excavation damage, and equipment failure for more than three decades. However, the key to learning from past incidents is to identify the underlying causes of the incidents, rather than direct causes. Existing approaches for utilizing past failure data are not able to fully capture the dependencies of both direct and underlying causes and their contribution to the pipeline failures. There is a need to understand the synergy among direct and underlying causes and their collective ability to affect the integrity of the pipeline. Although a wealth of empirical knowledge has been accumulated from public incident databases and incident investigation reports, learning is still limited due to the high expense of manual content analysis and lack of appropriate techniques to gain insights from past incidents. This research is aiming using artificial intelligence for understanding causal relations for direct as well as underlying causes behind the incidents. Natural language processing (NLP) techniques, both unsupervised and semi-supervised learning algorithms, are being employed to enable automated content analysis and extract causal relations articulated in the incident reports. The techniques are capable of identifying dependencies among the direct causes, underlying causes and other contributing factors behind the incidents. An artificial neural network (ANN) based causal reasoning model is also being developed utilizing all relevant factors for pipeline failure reported in the pipeline incident database and elsewhere, and causal relationships identified by the NLP technique. The advantage of using ANN is that it does not oversimplify the causal relationships leading to identification of both linear and non-linear components of causality and its non-imposition of restriction on the input variables (such as their distribution). The combination of NLP and ANN models will improve the ability to make full use of vast data resources of pipeline incidents.

620389 Improving Operations with Digital Twins

Abstract:

ABSTRACT for AIChE 2021 Spring Meeting

Improving Operations with Digital Twins

Joel Berg & James Cross
The MathWorks, Inc.

jberg@mathworks.com
508-330-7269

jcross@mathworks.com
617-605-5818

During the process engineering design phase, models are used to create equipment requirements, analyze operational performance, set control setpoints and protocols, and pursue economic goals. These models, often created using *process simulation* tools, are based on first principles, experimental data, or a combination (hybrid).

After plant construction, the limitations of the design model for representing the actual plant in operation become apparent. The design model is replaced by detailed controls protocols, with the actual system *state* being defined by sensor readings and other measurements. These data are or can be used to build a *digital twin*.

A digital twin is a model of a real system in operation, built using or calibrated by current data. These types of models represent the as-built system and inherently include effects of aging, deterioration, and failure phenomena. They are a valuable tool for interrogating what-if questions and ensuring safe and economic plant operation.

There are three main classes of decisions that are analyzed using digital twins:

- Process optimization. Should the control setpoints be changed?
- Predictive maintenance. When should plant maintenance be performed?
- Anomaly detection. Should the plant be shut down to avoid a catastrophic event?

In this talk we illustrate how the digital twin is created and utilized to answer questions like these, focusing our attention on a specific piece of process equipment. We also describe the value of the design model for augmenting failure modes and effects analyses, specifically for determining and quantifying fault signatures.

- Abstract Text

620388 Autonomous Process Model Identification

Abstract:

Autonomous Process Model Identification

Human errors are significant contributors to unavailability, waste, and undesirable environmental impact for industrial production. Avoiding errors and making better decisions is key for delivering superior operational outcomes. Decisions about industrial assets involve multiple dimensions such as operations, maintenance, and process improvements and these decisions can rely today on digital twins, which bring together the different aspects of industrial processes. However, the physical reality represented by these digital twins are subject to continuous change. Catalysts in reactors degrade, compressors and heat exchangers experience fouling, and gears and other mechanical parts wear out. Having access to data from the field could help to track some of these changes but updating the mathematical models underlying the digital twins is usually a demanding step. The scale of industrial production systems and the complexity of the involved processes and their interactions make fundamental first principles modeling methods prohibitive as the constant reconfiguration of the models require the attention of domain experts and although model parametrization and the online estimation of these parameters have been successful for applications with a limited scope, increasing their coverage to large scale systems can quickly become a bottleneck. Novel developments in the field of artificial intelligence bring up opportunities to tackle the modeling and model updating challenges by taking over some of the tasks and responsibilities of domain experts and help them scale their expertise.

In this presentation we will look specifically at the case of dynamic process model identification and how recurrent neural networks (RNN) could be used as building blocks for creating a self-learning system that can generate dynamic models for previously unseen systems. An analysis for the requirements of such a solution will be given and a general architecture with several algorithmic components will be proposed. The challenges with regard to some of the involved tasks – especially concerning hyperparameter optimization for the RNN models – will be discussed in detail and some alternatives for tackling these challenges will be presented. A prototype of the presented concept will be demonstrated in a case study of a digital twin for a continuous stirred tank reactor. The case study will cover a scenario, where the reactor experiences a disturbance leading to a reduction of the main reaction rate and where the autonomous learning system can detect this change and build a new model to recover prediction accuracy. The presentation will conclude with an overview of further challenges in realizing such autonomous self-learning systems for process model identification and the possible paths for their further development.

- Abstract Text

619157 The Digital Reactor – Using a Process Twin to Ensure Safe, Operable and Optimized Performance

Abstract:

Introduction

We are living through a digital revolution, often referred to as 'Industry 4.0', where production operations are being connected to smart digital technologies such as machine learning, artificial intelligence and big data. The drive is to connect these elements and derive real-time insights across the full spectrum of operations. The timing of this revolution coincides with advances in supportive technologies such as data science, where more data of higher quality is being collected from a wider range of sources and computation, where more processing power is available at significantly lower cost. Data analytics also helps, with rapidly evolving technologies in machine leaning, artificial intelligence and data mining.

In applying Industry 4.0 principles, there is a push to capture knowledge through experience and data at each step of the process lifecycle. For this paper, we introduce how such thinking affects our approach to the ultimate safe and optimal operation of industrial reactors and in particular, the steam-methane reformer. The focus is model-based – using a 'high-fidelity' model which can be developed throughout the cycle and which ultimately embodies the deep process knowledge which it accumulates along the way, including elements from reactor development, scale-up, design and operation. As new information becomes available, the model will be updated accordingly, preserving the knowledge within the model itself. The ultimate goal is to ensure the final reactor as installed is safe and exhibits maximum profitability for the operator.

Reactor model development

To satisfy the objective of a digital approach to reactor development, a systematic approach is followed to develop the reactor model. The preferred route is to utilize a high-fidelity model which completely describes the behaviour of the reaction. The useful modelling capabilities that are required include:

- ability to execute steady-state and dynamic simulation and optimization
- parameter estimation and data analysis capabilities to help determine accurate reaction kinetic parameters and heat-transfer coefficients from experimental and pilot data
- results management and visualization capabilities.

With a validated reaction kinetics model established from laboratory and/or pilot trials, further layers of detail can be added to cover the effects of catalyst loading, design arrangement, heat-transfer properties and pressure drops.

At the end of the reactor model development, the Digital Reactor model now contains enough knowledge to predict reaction and thermal performance for all anticipated reactor operating conditions. The model can now be used to optimize the operation of the commercial-scale reactor.

Steam-methane reforming (SMR) process

A detailed equation-based model has been built for a small steam-methane reforming process generating hydrogen for use in transportation. Such small-scale reactors are predicted to grow significantly in number as the hydrogen economy grows. The performance of the whole SMR process is described in detail, from catalyst behaviour to heat-transfer. The model has been tuned and validated against actual plant operating data so that it can be used in a digital twin role.

Reactor Operations

The SMR model has sufficient fidelity and predictive accuracy to be used on-line. It contains all the deep knowledge, developed through tuning against actual plant performance. The model can be deployed online, interfaced to plant data systems, to be used for the following applications:

- **Data reconciliation.** Plant data measurements are subject to drift and error. A confidence level can be applied to each plant measurement and the Digital Reactor model used to establish the 'best-fit' heat and material balance to the plant data.
- **Soft-sensing.** Many difficult or impossible-to-measure key performance indicators such as temperature and composition profiles within the reactor beds can instead be calculated using the model. Exposing such soft-sensed information leads to better decision-making such as improved control and enhanced safety.
- **Long-term health monitoring**. By combining a predictive model with live plant data, the model can be used to track and quantify drift in key process parameters such as catalyst activity.
- **Real-time optimization**. The Digital Reactor model can be used to maximize profit in real time, taking into account equipment and product quality constraints as well as current product demands, product prices, utility costs and feedstock costs.
- **Run-length prediction**. The Digital Reactor model can sometimes include forecasting to predict, for example, end-of-run dates when a catalyst bed may need regenerating or replacing.

To be able to achieve these modelling objectives, a standard IT solution framework is required which allows the model to run robustly on-line. A Digital Applications Platform (DAP) is selected with the following elements:

- **Execution Schedule Manager** controls how the platform operates, manages the timing and execution of the components listed below and controls the way in which the system recovers and corrects for errors
- **External Data Manager** handles communications and collects information from external data historians, distributed control systems and other plant databases to feed to the data validation module
- **Data Validation Module** applies extensive checking of all data read from the external servers and applies any remedial action to correct missing or invalid data. This module also applies validation rules to any results calculated by the models in the platform.
- **Computational Modules** are wrappers for the Digital Reactor and other models which perform simulations with a discrete focus such as those outlined above (from data reconciliation onwards).

Typically, the DAP is mounted onto a physical or virtual computer located somewhere on the plant network. The plant data, either from the Distributed Control System or the data historian, are read by the DAP. Once validated, the data are passed to the Digital Reactor model where the various modelling objectives outlined above can be carried out. The results of the Digital Reactor model can be used in 2 ways:

- **On-line closed-loop control.** Some of the calculated results will be sent back to the DCS or data historian and the plant control system will take over, adjusting the plant set-points to reflect the calculated optimum settings
- **On-line decision support.** The results will be presented to the operator in an easily readable format and the operator will then have the choice to implement the results or not. Generally, this is the preferred choice, especially early on when the model needs to be tested to ensure it is giving the right information.

In the system used by the authors, the validated plant data and Digital Reactor calculation results are stored in a SQL database. This allows the results to be more easily made available to operators and it is compliant with Industry 4.0 as its contents are available for use by any other App residing on the plant network. The information for engineers and operators is most efficiently provided via dashboards which employ graphics and easy-to-view operator actions, advising the operator where the plant is now and where it should be in the future.

Conclusions

The key take-aways from this paper are:

- Progress in digital technologies such as computation, data science and analytics are driving the digitalization of our approaches. The key emphasis is to use data to embed deep process knowledge into our models and to allow models to 'learn' as they are used to predict process performance
- Digital models should be developed for the whole process lifecycle, from R&D through to operations, to capture knowledge and always be using the full extent of this knowledge. The underlying software platform in which the Digital Reactor model is to be built should be carefully selected so as to have as much functionality as possible across the lifecycle
- A Digital Applications Platform (DAP) has been described which enables a reactor model to be run on-line as part of a standard approach. The DAP allows the model to access plant data, to allow it to be tuned and therefore maximizes its predictive accuracy. As a result, the model can be used as part of an advanced process control system or as an operations advisor in a decision support tool. In the latter, visualization is very important, to ensure the users' needs are satisfied.

A real-life small-scale industrial SMR process has been used to demonstrate the techniques.

620336 Topological Data Analysis for Multivariate Process Monitoring: Inspirations from Neuroscience

Abstract:

Process monitoring is essential in detecting abnormal behavior that might compromise efficiency and safety but is challenging due to the presence of multivariable interdependencies and of time-varying and nonlinear behavior [1,2]. A scientific field that requires monitoring a system with similar characteristics is neuroscience [3]. Specifically, the study of the brain through methods such as functional magnetic resonance imaging (fMRI) requires the analysis of high-dimensional multivariate time series that capture the activity and interdependencies of different areas of the brain in order to understand responses to different stimuli or to detect abnormal behavior or disease [4, 5]. Thus, there is a diverse set of insights and methods from neuroscience that can potentially be applied to process monitoring.

In this talk, we discuss connections found between the areas of neuroscience and multivariate process monitoring and how this leads to the incorporation of new data analysis techniques from topology and algebraic geometry. Prevalent data analysis methods in the process monitoring literature are based upon statistical techniques such as principal component analysis (PCA) and partial least squares (PLS) [6]. These methods are effective at identifying informational features from data (e.g., principal components) but might fail to identify other types of hidden features. For instance, data objects live in domains that can be described using topological and geometrical features; these features are robust to noise, generalize to high dimensions, and do not require statistical assumptions such as isotropy and stationarity. Here, we focus on the application of Riemannian geometry and algebraic topology to characterize datasets arising in multivariate process monitoring [7, 8]. Specifically, our approach analyzes process behavior by characterizing the geometrical structure of correlation matrices. We illustrate the benefits of these techniques using datasets from the Tennessee Eastman chemical process [9].

[1] Youqing Wang, Yabin Si, Biao Huang, and Zhijiang Lou. Survey on the theoretical research and engineering applications of multivariate statistics process monitoring algorithms: 2008–2017. *The Canadian Journal of Chemical Engineering*, 96(10):2073–2085, 2018.

[2] Thuntee Sukchotrat, Seoung Bum Kim, and Fugee Tsung. One-class classification-based control charts for multivariate process monitoring. *IIE transactions*, 42(2):107–120, 2009.

[3] Jonathan D Power, Alexander L Cohen, Steven M Nelson, Gagan S Wig, Kelly Anne Barnes, Jessica A Church, Alecia C Vogel, Timothy O Laumann, Fran M Miezin, Bradley L Schlaggar, et al. Functional network organization of the human brain. *Neuron*, 72(4):665–678, 2011.

[4] Nicolas Langer, Andreas Pedroni, Lorena RR Gianotti, Jurgen Hanggi, Daria Knoch, and Lutz J¨ancke. Functional brain network efficiency predicts intelligence. *Human brain mapping*, 33(6):1393–1406, 2012.

[5] Sophie Achard, Raymond Salvador, Brandon Whitcher, John Suckling, and ED Bullmore. A resilient, low-frequency, small-world human brain functional network with highly connected association cortical hubs. *Journal of Neuroscience*, 26(1):63–72, 2006.

[6] Salvador Garcıa-Munoz, Theodora Kourti, John F MacGregor, Arthur G Mateos, and Gerald

Murphy. Troubleshooting of an industrial batch process using multivariate methods. *Industrial & engineering chemistry research*, 42(15):3592–3601, 2003.

[7] Moo K Chung, Hyekyoung Lee, Alex DiChristofano, Hernando Ombao, and Victor Solo. Exact topological inference of the resting-state brain networks in twins. *Network Neuroscience*, 3(3):674–694, 2019.

[8] Marco Congedo, Alexandre Barachant, and Rajendra Bhatia. Riemannian geometry for eeg-based brain-computer interfaces; a primer and a review. *Brain-Computer Interfaces*, 4(3):155–174, 2017.

[9] Philip R Lyman and Christos Georgakis. Plant-wide control of the tennessee eastman problem. *Computers & chemical engineering*, 19(3):321–331, 1995.

620334 Applying Reinforcement Learning for Batch Trajectory Optimization in an Industrial Chemical Process

Abstract:

Reinforcement Learning (RL) is one of the three basic machine learning paradigms, alongside supervised and unsupervised learning. RL focuses on training an agent to learn an optimal policy, maximizing cumulative rewards from the environment of interest [1]. The recent developments in model-free RL have achieved remarkable success in various process optimization and control tasks, where multiple applications have been reported in the literature, including parameter tuning for existent single PID control loops [2], supply chain management [3] and robotics operations [4].

There are multiple challenges when applying RL in an industrial setting, but the main one concerns the training of the agent. In the learning phase, the agent estimates and improves its policy and value functions through a large number of trial and error iterations. Many input-output experimentations are required, which is obviously not feasible in an industrial chemical plant. As an alternative, a model of the plant can be utilized for training the agent and provide the input-output data. Both first principles and data-driven models are suitable, and both options are explored in this work.

In this work, we test two state-of-the-art RL approaches to optimize an industrial batch case study: Proximal Policy Optimization (PPO), Soft Actor Critic (SAC) and Advantage Actor Critic (A2C). These RL methods optimize the batch process by controlling the reaction conditions and maximizing the total reward (the reward is defined as the profit margin, subject to certain process and safety constraints). The batch optimal trajectories are compared in two scenarios. The first scenario uses, as an environment, a first principles model for training the agent. In the second scenario, a surrogate Long-Short-Term-Memory (LSTM) model is utilized, which combines both historical data from the reactor's operation and the first principle model estimates. The use of the LSTM is motivated by the fact that it helps mitigate accuracy issues from the first principle model by relying on the relationships found in the plant data. The optimized trajectories were compared to the current trajectories, and the RL optimal batch profiles show a 3% increase in product profit.

References

- Mnih, V., Kavukcuoglu, K., Silver, D., Graves, A., Antonoglou, I., Wierstra, D., & Riedmiller, M. (2013). Playing atari with deep reinforcement learning. arXiv preprint arXiv:1312.5602.
- Badgwell, T. A., Liu, K. H., Subrahmanya, N. A., & Kovalski, M. H. (2019). U.S. Patent Application No. 16/218,650.
- Gokhale, A., Trasikar, C., Shah, A., Hegde, A., & Naik, S. R. (2021). A Reinforcement Learning Approach to Inventory Management. In Advances in Artificial Intelligence and Data Engineering (pp. 281297). Springer, Singapore.
- Haarnoja, T., Zhou, A., Abbeel, P., & Levine, S. (2018). Soft actor-critic: Off-policy maximum entropy deep reinforcement learning with a stochastic actor. arXiv preprint arXiv:1801.01290.

620028 Leveraging Digital Image Data for Real-Time Inspection & Quality Prediction

Abstract:

As the price of machine vision hardware becomes more affordable, many manufacturers are seeking to leverage digital image data to achieve a variety of analytics and automation goals.

This presentation will discuss trends, advancements and considerations for effectively implementing color, thermal, 3D and polarized imaging to intelligently capture a variety of visual and non-visual product and process characteristics. Algorithms for calculating advanced image features and combining these features to predict key metrics will also be discussed.

Examples across a variety of industries will be provided to illustrate real-time monitoring and soft-sensor quality prediction accomplished through machine vision.

618856 How Do You Trust Your Asset and Process Safety Information If You Don't Frequently Measure Its Quality?

Abstract:

Asset and process safety information is the foundation of key safety, operations, engineering, and financial decisions we make every day. When that information is incomplete, inaccurate, inconsistent, or out of date, we pay for it in terms of higher operational and process risk, higher costs, inefficiencies, and reduced regulatory compliance.

Given the above premise, you would expect the majority of owner operators to formally measure and track the quality of their asset and process safety information and objectively know the true state of their information. After all, as Lord Kelvin said: "if you don't measure it, you cannot manage it." Unfortunately, that is not the case. Our research shows that Over 70% of plants have between 33% to 50%+ of their asset and process safety information missing, incomplete, inconsistent, or out of date with no formal quality management programs.

In this presentation, we will discuss:

- The need for formal quality management processes to frequently measure and report the true quality of their asset and process safety information.
- Underlying cultural, systems-related, and skills-related issues preventing companies from measuring and reporting quality metrics.
- Examples of what to measure and how to communicate it to stakeholders to drive positive change.
- The business benefits for companies that get to do it well.

620247 Measurements in Optimization Models

Abstract:

Online optimizer models use different types of measurements, like pressure, temperature, composition, valve position, flow to adjust internal parameters in the first case, where the objective is to update the model with present plant conditions. These parameters are kept constant during the second case where the optimum conditions are sought to maximize an objective function, it means that new plant conditions are predicted by the model.

The quality of the measurements defines the quality of the parameters and consequently, the optimizer model prediction. Even having a good measurement maintenance system, they can fail; and it is even worse if the quality of the measurement is not well supported.

During model execution, it is important to detect the most important parameters, those that ensure model predicts plant changes with acceptable performance. Then, model is designed including alternative measurements to keep it well-adjusted if the primary measure fails.

It is also necessary to incorporate a method to detect "online" when a measurement fails. There should be different ways to capture the scenarios, sometimes instrument quality detector is not enough, or validity range continues being correct although is not good. A continued tracking and monitoring of measurements quality helps to identify model opportunities and gaps.

Finally, if model is running online and an important measurement fails, a code identifies the situation and switches the model to use the secondary measurement to adjust the parameter. Making a good tuning of that, model will remain reasonably well parameterized and the optimal condition will be predicted well.

620937 Semi-Rigorous Models for Virtual Analyzers in Distillation Units

Abstract:

Model predictive multivariable control typically require proper measure for stream quality, either in terms of product composition or distillation curve point. Different technologies for online analysers are available and can provide quality measure with sufficient sampling frequency to be directly included in model predictive controllers. Nonetheless, analyser is generally expensive equipment and frequently not available on some important process streams, which relies on low frequency laboratory sampling. In such cases, several different methodologies for building inferential/virtual analyser have been proposed and used in the chemical industry, and particularly applied to the distillation processes, basing on available instrumentation. This paper describes semi-rigorous methodology applied to simple distillation column and atmospheric fractionator (as example, in refining industry, crude units, main fractionator after FCC reactor, etc).

The starting point is availability of a digital twin of the unit. From the digital twin, we can extract chemical and physical parameters necessary to build a set of simplified core equations for the virtual analyser. The simplified equation set is built in with some important characteristics:

- Can be solved only with the available process measurements (temperatures, pressures, flows, etc.)
- Can be solved high frequency, and with no risk of not converging to a solution, i.e. continuous and reliable predictions
- When running online, can provide a simple mechanism for reconciliation and feedback from available quality measures (e.g. lab sampling)

Key benefit from the methodology it that, being based on digital twin, the virtual analyser is relatively quick to be deployed and does not require big training datasets for initial data reconciliation. The paper will include some implemented case studies and their results.

633326 Sponsored Technology Workshop - Using Engineering Platforms to Understand the Cross Discipline Effect of Sustainability Research - Knovel

Abstract:

How to foster your research across the various engineering platforms. Engineering Village for journal and conference proceeding analysis of the research, Knovel for applied engineering to connect your findings across topics such as renewable energy or water resources, and Geofacets for an environmental perspective utilizing a map based interface.

620330 Industrial Threat Landscape Overview in the Chemical Industry

Abstract:

This presentation will present an overview of the industrial (OT/ICS/PCN) cyber threat landscape focusing on the relevant and current threat groups, incidents, and recommendations. The presentation will also explore the interconnection of cybersecurity and safety to help participants drive discussions with internal teams.

619991 Assessing Cybersecurity Risk in Industrial Internet of Things (IIoT) Applications

Abstract:

The advent of the Industrial Internet of Things (IIoT) within the chemical industry has the potential to expose control and safety systems to significant cyber security threats. These include:

- Exposure of the ICS to untrusted external connectivity
- Poor controls on generic IIoT Gateways or Edge Devices
- Weak Cloud Security
- Control conflicts and process upset derived from the advent multiple control schemes

This presentation will review the results of a cybersecurity risk assessment of 4 common IIoT applications and recommend secure architectures and associated security controls appropriate to each.

620332 Mitigating Threats and Malware through Enhanced Awareness, Detection & Response

Abstract:

Malware, ransomware, and cybersecurity related attacks against chemical and manufacturing companies are significantly on the rise. A good question to consider is – if an adversary wanted to gain access to a proprietary formula (for chemicals), or perhaps a manufacturing process, what defenses are in place to prevent that from happening, or alternatively detect that activity. Many companies are planning to place investments into cyber resiliency (risk mitigation) for the business in 2021 and beyond, but many are left wondering where to start? This session will provide an overview of key topics and considerations for non-security experts, so that companies are able to gain greater awareness when they are being targeted, and so they can enhance detection and improve overall preparedness for when bad things happen.

620333 Top 3 Mistakes When Protecting Your SIS from Cyber-Attack

Abstract:

We all know that cybersecurity threats are on the rise. With the increase in threat vectors as many companies relocate large numbers of centralized workers to remote workers this year, malware, ransomeware, phishing attempts and nation-state attacks are a clear and present danger to the control systems that we rely on. But, what about the Safety Instrumented System? The SIS is the last line of defense when other protections fail. Are you taking adequate precautions? Join us for a brief discussion on three common mistakes to consider when evaluation the cyber-safety of you safety system.

620195 "Digital Transformation 4.0" - Production Optimization with AI

Abstract:

With the constant pressure to increase plant efficiency and uptime in a highly competitive environment, condition-based maintenance and the optimization of the production processes become more and more important for a plant's economic success.

Siemens Predictive Analytics (SiePA) is a practical and robust tool that integrates human experience/ know-how and machine learning capability to extract the relevant information on the equipment and process status.

The information is summarized in an intuitive graphical user interface, used mainly by plant operators and maintenance staff. Thanks to AI, process data from the plant can be used optimally: Not only to predict the risk of a process failure and to display warnings, but also to efficiently diagnose possible malfunctions. The best prerequisites for increasing the efficiency, availability and safety of operations to a level that would have been unthinkable just a short time ago.

620194 Unlocking the Potential of AI: How to Implement AI-Based Predictive Analytics for Process Safety

Abstract:

Many companies across the industrial sectors are hiring data scientists and vendors to help harness the latest in AI to provide predictions and insights about pending maintenance for their assets. However, implementing useable solutions with AI at its core is more challenging than traditional expert based systems and software. In this presentation, Jim Eskew from SparkCognition will discuss what AI predictive analytics models are, how they are built and how they differ from expert models, and give a guide to the key elements to successfully deploy AI-based solutions. As part of his presentation, Jim will demo select functionality developed by SparkCognition which has been deployed and used by operators to reduce production disruptions and mitigate safety risks.

AUTHOR INDEX